MIXTURE MODEL-BASED CLASSIFICATION

MIXTURE MODEL-BASED CLASSIFICATION

Paul D. McNicholas

McMaster University
Ontario, Canada

CRC Press
Taylor & Francis Group
Boca Raton London New York

CRC Press is an imprint of the
Taylor & Francis Group, an **informa** business

A CHAPMAN & HALL BOOK

CRC Press
Taylor & Francis Group
6000 Broken Sound Parkway NW, Suite 300
Boca Raton, FL 33487-2742

First issued in paperback 2020

© 2017 by Taylor & Francis Group, LLC
CRC Press is an imprint of Taylor & Francis Group, an Informa business

No claim to original U.S. Government works

ISBN-13: 978-1-4822-2566-2 (hbk)
ISBN-13: 978-0-367-73695-8 (pbk)

Library of Congress Cataloging-in-Publication Data

Names: McNicholas, Paul D., author.
Title: Mixture model-based classification / Paul D. McNicholas.
Description: Boca Raton : Taylor & Francis, 2016.
Identifiers: LCCN 2016010443 | ISBN 9781482225662 (alk. paper)
Subjects: LCSH: Multiple comparisons (Statistics) |
Classification--Methods--Mathematics. | Discriminant analysis. | Mixture
distributions (Probability theory) | Hierarchical clustering (Cluster
analysis)
Classification: LCC QA278.4 .M45 2016 | DDC 519.5/35--dc23
LC record available at https://lccn.loc.gov/2016010443

Visit the Taylor & Francis Web site at
http://www.taylorandfrancis.com

and the CRC Press Web site at
http://www.crcpress.com

For my family:
Sharon, Oscar, Constance, and Henry

and in loving memory of my mother:
Patricia

Contents

List of Figures

List of Tables

Preface

Mixture models have been used for classification and clustering for over 50 years, and the topic has received increasing attention since the end of the last century. In fact, these mixture model-based approaches have burgeoned into an important subfield of classification. The volume of work on the subject and the breadth of its application justify its treatment in a monograph. However, the genesis of this book lies in my experiences introducing the topic to new Ph.D. students. The task of starting a Ph.D. student on a course of research on model-based approaches to clustering and classification previously involved assigning much, often cumbersome, reading material. This reading material consists of various media, mainly journal articles and book chapters, that use different approaches to make generally similar points. The different notation and arguments do not present a major problem in themselves; however, there is a distinct lack of consistency on some crucial points and this is difficult for a starting Ph.D. student. The origin of mixture models as a tool for clustering was not clearly understood, which deprived students of a natural beginning in understanding why this is a good idea. This matter is finally resolved at the beginning of Chapter 2 so that future students can understand that the initial question concerned how a cluster should be defined; it was ten years later that what we now call model-based clustering was first executed. A copy of a key text was provided to me by Lizette Royer Barton, a reference archivist at the Drs. Nicholas and Dorothy Cummings Center for the History of Psychology, University of Akron, Ohio, and I am very grateful to her in this regard. I am also most grateful to John Wolfe for providing a copy of his famous but also elusive master's thesis, from which I was able to get to the work of Tiedeman (1955) and others.

This monograph is laid out to progress naturally. Broadly, the three chapters immediately following the introduction cover Gaussian mixtures, mixtures of factor analyzers and extensions thereof, variable selection, and high-dimensional applications. Then the departure from the Gaussian assumption begins with consideration of mixtures of distributions that parameterize concentration, or tail weight. This departure continues with mixtures of skewed distributions, mixtures of distributions that parameterize both skewness and concentration, and mixtures of multiple scaled distributions. The remaining chapters mostly cover special topics, so to speak, including methods for clustering and classification of longitudinal data, cluster-weighted models, averaging mixture models, and so on. Notable exceptions are Section 9.1, wherein the definition of a cluster is explored, and Section 9.2, where the existence of a "best" clustering and classification method is discussed. While the layout is intended to assist a beginning Ph.D. student, it will equally benefit anyone who is coming to the subject anew or afresh. More seasoned readers will find the chapter and section titles self-explanatory. While its genesis lies in experiences gained through supervising Ph.D. students, this is a book for anyone with an interest in, or a use for, mixture model-based approaches to classification and clustering. The question of how much mathematical detail to include is addressed in a practical manner: where including mathematical detail is considered helpful in achieving a broad understanding, it is included; otherwise, the reader is directed elsewhere. Moreover, there are already several very good books on mixture models (e.g., Everitt and Hand, 1981; Titterington et al., 1985; McLachlan and Peel, 2000a; Frühwirth–Schnatter, 2006) and this monograph does not seek to replicate their contents.

I am most grateful to David Grubbs of the Taylor & Francis Group for his suggestion, back in 2011, that I write a monograph and for his unerring patience as I repeatedly missed deadlines. I am thankful to Dr. Brendan Murphy for starting me on my ongoing journey through the field. A number of my current and former supervisees kindly provided comments on parts of this book, while others helped with figures and analyses. In this regard, I am thankful for the assistance of Dr. Jeffrey Andrews, Dr. Ryan Browne, Dr. Utkarsh Dang, Dr. Sanjeena Dang, Dr. Matthieu Marbac, Paula Murray, Yang Tang, Dr. Cristina Tortora, and Dr. Irene Vrbik. I am particularly indebted to Dr. Brian Franczak for his invaluable assistance with the figures and analyses in Chapters 6 and 7. I am most grateful to Dr. Douglas Steinley, who read a draft of this monograph and provided some very helpful comments. I also wish to thank Drs. Salvatore Ingrassia and Antonio Punzo, who provided very helpful feedback on parts of this monograph.

I have been very fortunate to work with several wonderful collaborators on real clustering problems; in this regard, I am particularly grateful to Dr. Chris Findlay and John Castura, both of Compusense Inc. I wish to express very special and heartfelt thanks to Professor Petros Florides, Eamonn Mullins, Dr. Donal O'Donovan, and Dr. Myra O'Regan, all either currently or formally of Trinity College Dublin, Ireland. Suffice it to say that without their respective interventions and kindnesses, I probably would not have completed even an undergraduate degree. A final word of thanks goes to my wife, Sharon; without her patience and support, this book would not have come to fruition.

Guelph   Paul D. McNicholas

1

Introduction

1.1 Classification

Broadly, classification is the process of assigning group membership labels to unlabelled observations. In this context, a group may be a class or a cluster. There are three species of classification: supervised, semi-supervised, and unsupervised. The level of supervision refers to the extent to which there is *a priori* knowledge of the labels and how such knowledge is used. In unsupervised classification, also known as clustering, no observations are *a priori* labelled or, at least, the data are treated as such. In the other two species, some observations are *a priori* labelled and this knowledge is used to infer labels for the unlabelled observations. Suppose that there are n observations and $k < n$ of them are labelled. Semi-supervised classification uses all n observations to infer, or learn, labels for the $n - k$ unlabelled observations. Supervised classification, also known as discriminant analysis, uses only the k labelled observations to infer labels for the $n - k$ unlabelled observations.

As Hartigan (1975) points out, classification techniques were originally developed for biological taxonomy. Many interesting exchanges about the rights and wrongs, and indeed the purpose, of classification occurred in the 1950s, 1960s, and 1970s. Past exchanges about the fundamentals of statistical inference, involving pioneers such as Fisher, DiFinetti, and Savage, have become well known (cf. Barnett, 1999). However, the robust and sometimes amusing exchanges that took place concerning classification are less famous. Hartigan (1975) gives an excellent account of work on classification prior to the mid-1970s. This includes several quotations from Cormack (1971), who provides an earlier and very interesting review of classification that ends with the following quote from the "brilliant, witty and utterly scathing discussion" of Johnson (1968a, 1970):

> Anyone who is prepared to learn quite a deal of matrix algebra, some classical mathematical statistics, some advanced geometry, a little set theory, perhaps a little information theory and graph theory, and some computer technique, and who has access to a good computer and enjoys mathematics... will probably find the development of new taximetric method much more rewarding, more up-to-date, more 'gen-

eral', and hence more prestigious than merely classifying animals or plants or working out their phylogenies.

The work of Cormack (1971) was read before the Royal Statistical Society and the discussion thereof is very interesting. Hartigan (1975, p. 8) quotes part of the following excerpt from the response of Gower (1971):

> No doubt much "numerical taxonomic" work is logically unsound but it has acquainted statisticians with some new problems whose separate identities are only now beginning to emerge. If statisticians do not like the formulations and solutions proposed they should do better, rather than denigrate what others have done. Taxonomists must find it infuriating that statisticians, having done so little to help them, laugh at their efforts. I hope taxonomists who have real and, I think, interesting problems find it equally funny that so much statistical work, although logically sound, and often mathematically complicated (and surely done for fun), has little or no relevance to practical problems. They might prefer imperfect solutions to ill-defined problems than perfect solutions to well-defined non-problems; at least they can hope for improvements. Perhaps Dr Cormack's review will stimulate more statisticians to take these problems seriously.

Indeed, there are quite a number of statisticians, and others, working in the field of classification today who take such problems very seriously. Perhaps this can be attributed, at least in some small part, to the review of Cormack (1971) and the discussion thereof, as well as the work of others around the same time. However, insofar as many statisticians are concerned, nothing focuses the mind like a real problem, and the last few decades have seen a rapid growth in the emergence of real classification problems. Such problems emanate from a wide variety of fields, including astrophysics, biology, food science, political science, psychology, and zoology. This rapid growth in real problems has been accompanied by extraordinary advances in computing power — at the turn of the century, multicore computers were uncommon; today, cellular phones are multicore. Of course, this technological leap is not independent of the explosion in data-driven classification problems. Together, however, they make for a fertile environment for classification research: no shortage of data-driven problems and very little in the way of computational limitations.

Rather than further delving into the very interesting history of classification here, the reader is referred to Hartigan (1975, Chapter I) as a starting point for his or her own investigations. The historical development of Gaussian mixture models as a tool for classification will be summarized in Section 2.1, which also includes some background and historical discussion about the definition of a cluster. In brief, the position taken in this monograph is that a cluster is defined as one (unimodal) component of an appropriate mixture model. A detailed argument in this direction is presented in Section 9.1.

1.2 Finite Mixture Models

A random vector $\mathbf{X}$ arises from a parametric finite mixture distribution if, for all $\mathbf{x} \subset \mathbf{X}$, its density can be written

$$f(\mathbf{x} \mid \boldsymbol{\vartheta}) = \sum_{g=1}^{G} \pi_g f_g(\mathbf{x} \mid \boldsymbol{\theta}_g), \qquad (1.1)$$

where $\pi_g > 0$, such that $\sum_{g=1}^{G} \pi_g = 1$, is the gth mixing proportion, $f_g(\mathbf{x} \mid \boldsymbol{\theta}_g)$ is the gth component density, and $\boldsymbol{\vartheta} = (\pi_1, \ldots, \pi_G, \boldsymbol{\theta}_1, \ldots, \boldsymbol{\theta}_G)$ is the vector of parameters. Note that $f(\mathbf{x} \mid \boldsymbol{\vartheta})$ in (1.1) is called a G-component finite mixture density. The component densities $f_1(\mathbf{x} \mid \boldsymbol{\theta}_1), f_2(\mathbf{x} \mid \boldsymbol{\theta}_2), \ldots, f_G(\mathbf{x} \mid \boldsymbol{\theta}_G)$ are often taken to be of the same type, i.e., $, f_g(\mathbf{x} \mid \boldsymbol{\theta}_g) = f(\mathbf{x} \mid \boldsymbol{\theta}_g)$ for all g. Extensive details on finite mixture models and their applications are given in the well-known texts by Everitt and Hand (1981), Titterington et al. (1985), McLachlan and Basford (1988), McLachlan and Peel (2000a), and Frühwirth–Schnatter (2006).

Historically, the Gaussian mixture model has been the most popular mixture model in classification applications. A Gaussian mixture model has density

$$f(\mathbf{x} \mid \boldsymbol{\vartheta}) = \sum_{g=1}^{G} \pi_g \phi(\mathbf{x} \mid \boldsymbol{\mu}_g, \boldsymbol{\Sigma}_g), \qquad (1.2)$$

where

$$\phi(\mathbf{x} \mid \boldsymbol{\mu}_g, \boldsymbol{\Sigma}_g) = \frac{1}{\sqrt{(2\pi)^p \mid \boldsymbol{\Sigma}_g \mid}} \exp\left\{ -\frac{1}{2}(\mathbf{x} - \boldsymbol{\mu}_g)' \boldsymbol{\Sigma}_g^{-1} (\mathbf{x} - \boldsymbol{\mu}_g) \right\}$$

is the density of a random variable $\mathbf{X}$ from a multivariate Gaussian distribution with mean $\boldsymbol{\mu}_g$ and covariance matrix $\boldsymbol{\Sigma}_g$. Note that $\phi_p(\mathbf{x} \mid \boldsymbol{\mu}_g, \boldsymbol{\Sigma}_g)$ will sometimes be written to emphasize that $\mathbf{X}$ is p-dimensional.

A family of mixture models arises when various constraints are imposed upon component densities, and most often upon the covariance structure; the result is a flexible modelling paradigm that incorporates more and less parsimonious models. For example, three straightforward constraints on component covariance matrices in a Gaussian mixture model are $\boldsymbol{\Sigma}_g = \sigma_g^2 \mathbf{I}_p$, $\boldsymbol{\Sigma}_g = \sigma^2 \mathbf{I}_p$, and $\boldsymbol{\Sigma}_g = \boldsymbol{\Sigma}$ (cf. Friedman and Rubin, 1967; Scott and Symons, 1971; Gordon, 1981). The three corresponding mixture models, together with the most general model (1.2), give a family of four Gaussian mixture models. In a typical application, one would fit all four models and select the best one via some criterion. A very famous family of Gaussian mixture models, model selection criteria, and other pertinent topics are discussed in Chapter 2.

1.3 Model-Based Clustering, Classification, and Discriminant Analysis

Broadly, the term "model-based clustering" is used to describe the process of clustering via a statistical model. However, it is often taken to mean the use of a mixture model for clustering or the use of a family of mixture models for clustering. Herein, the term "model-based clustering" is used in the mixture model context. Similarly, "model-based classification" (e.g., McNicholas, 2010) or "partial classification" (cf. McLachlan, 1992, Section 2.7) is the use of mixture models for semi-supervised classification, and "model-based discriminant analysis" (Hastie and Tibshirani, 1996) is the use of mixture models for supervised classification.

Model-based clustering, classification, and discriminant analysis are best illustrated through their respective likelihoods. Note that, while a Gaussian mixture model is used for illustration in the following, an analogous approach applies for other component densities. Let $\mathbf{z}_i = (z_{i1}, \ldots, z_{iG})$ denote the component membership of observation i, so that $z_{ig} = 1$ if observation i belongs to component g and $z_{ig} = 0$ otherwise. Consider a clustering scenario, where n p-dimensional data vectors $\mathbf{x}_1, \ldots, \mathbf{x}_n$ are observed and all are unlabelled or treated as unlabelled. Then the Gaussian model-based clustering likelihood can be written

$$\mathcal{L}(\boldsymbol{\vartheta}) = \prod_{i=1}^{n} \sum_{g=1}^{G} \pi_g \phi(\mathbf{x}_i \mid \boldsymbol{\mu}_g, \boldsymbol{\Sigma}_g).$$

Note that $\mathbf{z}_i$ is considered a realization of $\mathbf{Z}_i$, which is a random variable that follows a multinomial distribution with one draw on G categories with probabilities given by $\pi_1, \ldots, \pi_G$. In fact, $\mathbf{Z}_1, \ldots, \mathbf{Z}_n$ are assumed independent and identically distributed according to a multinomial distribution with one draw on G categories with probabilities $\pi_1, \ldots, \pi_G$. As McLachlan and Peel (2000a, Section 1.9) point out, the mixing proportion π_g can be interpreted as the *a priori* probability that an observation $\mathbf{x}_i$ belongs to component g. The corresponding *a posteriori* probability is

$$\mathbb{P}[Z_{ig} = 1 \mid \mathbf{x}_i] = \frac{\pi_g \phi(\mathbf{x}_i \mid \boldsymbol{\mu}_g, \boldsymbol{\Sigma}_g)}{\sum_{h=1}^{G} \pi_h \phi(\mathbf{x}_i \mid \boldsymbol{\mu}_h, \boldsymbol{\Sigma}_h)}. \tag{1.3}$$

Note that the *a posteriori* expected value $\mathbb{E}[Z_{ig} \mid \mathbf{x}_i]$ is also given by (1.3), i.e., $\mathbb{E}[Z_{ig} \mid \mathbf{x}_i] = \mathbb{P}[Z_{ig} = 1 \mid \mathbf{x}_i]$.

After the parameters have been estimated, the predicted classifications are given by (1.3). For convenience, write

$$\hat{z}_{ig} := \frac{\hat{\pi}_g \phi(\mathbf{x}_i \mid \hat{\boldsymbol{\mu}}_g, \hat{\boldsymbol{\Sigma}}_g)}{\sum_{h=1}^{G} \hat{\pi}_h \phi(\mathbf{x}_i \mid \hat{\boldsymbol{\mu}}_h, \hat{\boldsymbol{\Sigma}}_h)}, \tag{1.4}$$

for $i = 1, \ldots, n$ and $g = 1, \ldots, G$. The fact that these *a posteriori* predicted classifications are soft — i.e., each observation has a probability of belonging to each component under the fitted model — is often considered an advantage of the mixture model-based approach. For example, in a $G = 2$ component scenario, it is useful to know whether or not the fifth observation is solidly within one component under the fitted model, e.g., whether $\mathbf{z}_5 = (0, 1)$ or $\mathbf{z}_5 = (0.47, 0.53)$. Having soft classifications can be very useful in practice, e.g., when interpreting results or comparing different clustering methods. However, in some applications it is desirable to harden the *a posteriori* classifications and the most popular way to do this is to report maximum *a posteriori* (MAP) classifications, i.e., MAP$\{\hat{z}_{ig}\}$, where

$$
\mathrm{MAP}\{\hat{z}_{ig}\} = \begin{cases} 1 & \text{if } g = \arg\max_h\{\hat{z}_{ih}\}, \\ 0 & \text{otherwise.} \end{cases}
$$

Next consider the model-based classification paradigm. Suppose that k of the n observations are labelled as belonging to one of G classes and order them, without loss of generality, so that the first k are labelled, i.e., $\mathbf{x}_1, \ldots, \mathbf{x}_k, \mathbf{x}_{k+1} \ldots \mathbf{x}_n$. Then the Gaussian model-based classification likelihood can be written

$$
\mathcal{L}(\vartheta) = \prod_{i=1}^{k} \prod_{g=1}^{G} \left[\pi_g \phi(\mathbf{x}_i \mid \boldsymbol{\mu}_g, \boldsymbol{\Sigma}_g)\right]^{z_{ig}} \prod_{j=k+1}^{n} \sum_{h=1}^{H} \pi_h \phi(\mathbf{x}_j \mid \boldsymbol{\mu}_h, \boldsymbol{\Sigma}_h), \tag{1.5}
$$

for $H \geq G$. Note that, while it is possible to fit a number of components H greater than the number of observed classes G, it is often assumed that $H = G$. The predicted classifications are given by $\hat{z}_{ig}$, for $i = k+1, \ldots, n$ and $g = 1, \ldots, G$; once again, these may be hardened to MAP classifications. From (1.5), model-based clustering can be considered a special case of model-based classification that arises upon setting $k = 0$ within the latter paradigm.

Finally, consider model-based discriminant analysis and suppose that each known class may correspond to multiple component densities in a mixture. First, a Gaussian mixture model is fitted to the labelled observations in each known class and the number of components is determined via some criterion (cf. Section 2.4). This step is effectively a model-based clustering for each known class so that $\mathcal{G}_g \geq 1$ components are fitted to the gth class. It is convenient to write $\mathcal{G} = \mathcal{G}_1 + \cdots + \mathcal{G}_G$. Then the likelihood

$$
\mathcal{L}(\vartheta) = \prod_{i=1}^{k} \prod_{g=1}^{\mathcal{G}} \left[\pi_g \phi(\mathbf{x}_i \mid \boldsymbol{\mu}_g, \boldsymbol{\Sigma}_g)\right]^{z_{ig}}
$$

is formed from the k labelled observations; using the maximum likelihood estimates arising from this likelihood, the expected values $\hat{z}_{ig}$ are then computed for $i = k+1, \ldots, n$ and $g = 1, \ldots, G$. These expected values play the role of a discriminant rule and the predicted classifications are based on the associated *a posteriori* classifications or the hardened analogues thereof. In either case, the correspondence between the components $\mathcal{G}_1, \ldots, \mathcal{G}_G$ and the classes must be accounted for when reporting results. Later within this monograph, e.g., Section 9.2, it will be argued that it is often desirable to force $\mathcal{G}_g = 1$ for all $g \in \{1, \ldots, G\}$, in which case the approach is essentially straightforward discriminant analysis.

1.4   Comparing Partitions

The need to compare partitions often arises in classification and clustering settings. For example, predicted and true classes may be compared or two alternative clustering results may be compared. In comparing partitions, it is useful to consider pair agreements. For notational convenience, it is useful to follow Steinley (2004) and consider a table of pairwise agreements and disagreemnts (Table 1.1).

Table 1.1
Cross-tabulation of pairs for two partitions, where rows represent pairs of observations from one partition and columns represent pairs from another partition.

	Same group	Different groups
Same group	A	B
Different groups	C	D

The Rand index (RI; Rand, 1971) is the ratio of the pair agreements to the total number of pairs or, using the notation in Table 1.1,

$$\text{RI} = \frac{A + D}{N}, \tag{1.6}$$

where $N = A + B + C + D$ is the total number of pairs. Clearly, $\text{RI} = 1$ corresponds to perfect class agreement. Morey and Agresti (1984) note that chance agreement will tend to lead to the value in (1.6) being inflated and they propose a correction to account for chance agreement. As Steinley (2004) points out, Hubert and Arabie (1985) found an error in the corrected index given by Morey and Agresti (1984) and they introduce their own adjusted Rand index, which will be denoted ARI hereafter. Using the notation in Table 1.1,

$$\text{ARI} = \frac{N(A + D) - [(A + B)(A + C) + (C + D)(B + D)]}{N^2 - [(A + B)(A + C) + (C + D)(B + D)]} \tag{1.7}$$

(cf. Steinley, 2004). The ARI, (1.7), has expected value 0 under random classification and takes a value of 1 for perfect class agreement. This is apparent from the general form of the correction, i.e.,

$$\text{corrected index} = \frac{\text{index} - \text{expected index}}{\text{maximum index} - \text{expected index}}.$$

Negative values of the ARI are also possible and can be interpreted as classifications that are worse than would be expected under random classification. Note that the ARI has no well-defined lower bound; Hubert and Arabie (1985) comment that "the required normalization would offer no practical benefits".

The ARI has become the most popular method for assessing class agreement in mixture model-based applications. However, the misclassification rate is sometimes used. Steinley (2004) gives detailed simulations showing that the ARI is preferable to the misclassification rate when the number of clusters equals the number of known classes; various favourable properties of the ARI are also discussed therein. Situations quite often arise when the number of clusters exceeds the number of classes; accordingly, comparing the ARI values and the misclassification rates for the data presented in Tables 1.2 and 1.3 is of practical interest. If multiple clusters are used to represent a class, then all but one of these clusters will be treated as misclassifications in the calculation of the misclassification rate. Accordingly, the misclassification rate associated with Table 1.2 is 0.26. Although Table 1.3 clearly depicts an inferior classification, the associated misclassification rate is also 0.26. The respective ARI values are much more sensible, with Table 1.2 (ARI = 0.61) emerging as being reflective of a better classification result than Table 1.3 (ARI = 0.29). Note that, although it performs very well in general, care should be taken when using the ARI for small values of n (see Section 7.7.3, for example).

Table 1.2
A sample cross-tabulation of true classes (1,2) against clusters (A–C).

	A	B	C
Class 1	48	42	10
Class 2	0	0	100

Table 1.3
Another sample cross-tabulation of true classes (1,2) against clusters (A–C).

	A	B	C
Class 1	48	10	42
Class 2	0	0	100

For completeness, it should be mentioned that several others have introduced indices that are based on pairwise agreement. As Hubert and Arabie

(1985) mention, Johnson (1968b), Mirkin and Chernyi (1970), and Arabie and Boorman (1973) adopt the index D/N, and Hubert (1977) proposes

$$\frac{A - D}{N}.$$

Steinley (2004) points out that the Jaccard index (Downton and Brennan, 1980) can be written

$$\frac{A}{A + B + C},$$

and that the Fowlkes–Mallows index (Fowlkes and Mallows, 1983) is given by

$$\frac{A}{\sqrt{(A + B)(A + C)}}.$$

1.5   R Packages

Several packages for the R software (R Core Team, 2015) are used in this monograph. Details of these packages are given in Table 1.4.

Table 1.4
The R packages used herein, with version number and relevant citations.

Name	Version	Relevant Citations
alr3	2.0.5	Weisberg (2005, 2010)
e1071	1.6-3	Meyer et al. (2014)
Flury	0.1-3	Flury (2012)
gclus	1.3.1	Hurley (2004)
longclust	1.2	McNicholas, Jampani, and Subedi (2015), McNicholas and Subedi (2012), McNicholas and Murphy (2010a)
mclust	4.3	Fraley et al. (2012, 2014), Fraley and Raftery (2002b), Celeux and Govaert (1995)
MixGHD	1.8	Tortora et al. (2015a,b, 2016), Browne and McNicholas (2015)
mixture	1.1	Browne and McNicholas (2014a,b), Celeux and Govaert (1995)
pgmm	1.2	McNicholas, ElSherbiny, McDaid, and Murphy (2015), McNicholas (2010), McNicholas and Murphy (2008, 2010b)
teigen	2.0.81	Andrews and McNicholas (2012, 2015)
vscc	0.2	Andrews and McNicholas (2013, 2014)

1.6 Datasets

The datasets used for illustration in this monograph are summarized in Table 1.5. Note that "UCI Repository" in Table 1.5 refers to the UCI Machine Learning Repository (Lichman, 2013) and "Website" refers to the author's website www.paulmcnicholas.info.

Table 1.5
The datasets used herein, with the number of samples, dimensionality (Vars.), number of classes (Cls.), and source.

Name	Samples	Vars.	Cls.	Source
ais	202	11	2	alr3
banknote	200	6	2	mclust
bankruptcy	66	2	2	MixGHD
body	507	24	2	gclus
coffee	43	12	2	pgmm
Colon Cancer	62	461	2	Website
crabs	200	5	4	MASS
diabetes	145	3	3	mclust
faithful	272	2	NA	datasets
f.voles	86	7	2	Flury
iris	150	5	3	datasets
Leukaemia	72	2,030	2	Website
olive	572	8	9	pgmm
Urinary Disease	120	6	2	UCI Repository
wine	178	27	3	pgmm
Wisconsin Breast Cancer	569	30	2	UCI Repository
x2	300	2	3	mixture

1.7 Outline of the Contents of This Monograph

The contents of this monograph proceed as follows. The first section of Chapter 2, i.e., Section 2.1, is devoted to the historical development of model-based clustering, and includes discussion on the genesis of model-based clustering and some historical context for the definition of a cluster. This material presents a bridge between the background material in this introduction and the remaining chapters. Chapters 2, 3, and 4 cover Gaussian mixtures, mixtures of factor analyzers and extensions thereof, variable selection, and high-dimensional applications. Then the departure from the Gaussian assumption begins with consideration of mixtures of distributions that parameterize concentration, or tail weight (Chapter 5). This departure continues with mixtures of skewed distributions and mixtures of distributions that parameterize both skewness and concentration, including mixtures of multiple scaled distributions (Chapters 6 and 7). Chapter 8 covers methods for clustering and classification of longitudinal data. Chapter 9 begins with a discussion on the definition of a cluster (Section 9.1) and the best approach to clustering, classification, and discriminant analysis (Section 9.2), and proceeds to cover miscellaneous special topics while also providing reflections on some of the major challenges in the field.

2

Mixtures of Multivariate Gaussian Distributions

2.1 Historical Development

The first instance of Gaussian mixture models being used for clustering is sometimes attributed to Wolfe (1963). However, this is not entirely accurate. Wolfe (1963) uses the idea of a mixture to define a cluster, or type:

> A type is a distribution which is one of the components of [a] mixture of distributions.

As Wolfe (1963) points out, this definition of type, or cluster, is similar to that used by Tiedeman (1955). The paper by Tiedeman (1955) is a prescient paper that builds on famous works by Pearson (1894) and Rao (1952) to point out, *inter alia*, the need for more than two components in classification and clustering applications. A driving force behind the work of Tiedeman (1955) is to encourage work on what we now know as clustering:

> I will describe the problem created when type is not specified beforehand because I would like to encourage interest in this problem in order that a complete and a consistent system of statistics might become available for the study of types with measurement variables.

Because the idea of defining a cluster in terms of a component in a mixture model goes back to Tiedeman (1955), it is worth noting exactly how he formulated the problem:

> Consider G observation matrices each of which generates a density function of the form given by equation [1]. Throw away the type identification of each observation set and you have a mixed series of unknown density form.

Of course, [1] is the density of a Gaussian random variable. The objective, as laid down by Tiedeman (1955), is then

> ...to solve the problem of reconstructing the G density functions of original types. Such is the problem; its solution is extremely difficult if possible.

Over the subsequent two decades, much energy was invested in its solution, led by Wolfe (1963, 1965).

The clustering procedures developed by Wolfe (1963) are not based on maximizing, or otherwise exploiting, the likelihood of a Gaussian mixture model. Of this clustering methodology, Wolfe (1963, p. 76) writes:

> The methods described in this thesis are not only bad, they have been rendered obsolete by the author's own subsequent work. By the time this thesis was 3/4 finished, another method called "maximum-likelihood analysis of types" was developed which utilizes all of the information in the sample with maximal efficiency, employs no arbitrary measures of distance, utilizes no arbitrary parameter such as a selection radius, generates no extraneous clusters, and provides confidence intervals, and significance tests.

The subsequent work referred to here is the paper by Wolfe (1965), which seems to be the first published example of (Gaussian) model-based clustering. The intention here is not at all to denigrate the thesis by Wolfe (1963), which stands as an extremely important piece of work in the history of mixture model-based approaches to classification and clustering. Rather, it is to clarify the respective contributions of different pieces of work.

Wolfe (1963, Chapter I.D) also discusses two alternative definitions of a cluster. This first alternative is:

> A type is a mode in a distribution.

Wolfe (1963) points out that this definition is similar to that used by Cattell (1957) and that, although not entirely inconsistent with the mixture component-based definition, it is less flexible because the component variance is not taken into account. The clustering approaches developed by Wolfe (1963) perhaps adhere more directly to this definition than to the component density-based one, albeit implicitly. The second alternative definition is:

> A type is a set of objects which are more similar to each other than they are to objects not members of the set.

Wolfe (1963) also quotes a related definition from Thorndike (1953):

> For a given value of k, how shall we assign N specimens to k categories so that the average of the within-categories distances will be a minimum?

Wolfe (1963) points out several problems with such definitions. Beyond the issues he raises, the fact that definitions such as the second alternative definition given by Wolfe (1963) are satisfied by a solution where each point is assigned to its own cluster is problematic. Mixture component density-based definitions, such as those used by Tiedeman (1955) and Wolfe (1963), are much cleaner and more satisfactory in general; see Section 9.1 for further discussion on the definition of a cluster. Note that some argue that it might also be desirable to consider situations in which a cluster can contain more than one component density, and this will be discussed further in Sections 2.5 and 9.1.

Wolfe (1965) computes maximum likelihood estimates for Gaussian model-based clustering. Software encompassing four different parameter estimation techniques is presented, including an iterative scheme. This software is effective for up to five variables and as many as six components. Day (1969) introduces an iterative technique for finding maximum likelihood estimates when the covariance matrices are held equal, and discusses clustering applications. Wolfe (1970) develops iterative approaches for finding maximum likelihood estimates in the cases of common and differing component covariance matrices, respectively, and illustrates these approaches for clustering. Interestingly, Wolfe (1970) draws an analogy between his approach for Gaussian mixtures with common covariance matrices and one of the criteria described by Friedman and Rubin (1967). Scott and Symons (1971) consider approaches and parameter estimation in a Gaussian model-based classification scenario, and point out the analogy of one of these approaches with the work of Edwards and Cavalli–Sforza (1965). Since its introduction, the expectation-maximization (EM) algorithm (Dempster et al., 1977) has been by far the most popular approach for mixture model parameter estimation. Broadly, the EM algorithm is an iterative procedure for finding maximum likelihood estimates when data are incomplete and, as Titterington et al. (1985, Section 4.3.2) point out, similar treatments had previously been employed by Baum et al. (1970), Orchard and Woodbury (1972), and Sundberg (1974).

In his monograph on classification, Gordon (1981) returns to these analogies and discusses several others (cf. Gordon, 1981, Section 3.5). At the beginning of their monograph, McLachlan and Basford (1988) give a detailed history on mixture models, including much on univariate mixtures, which are all but omitted from the present monograph. Banfield and Raftery (1993) consider eigen-decompositions of the component covariance matrices and study several resulting models. Celeux and Govaert (1995) build on these, resulting in a family of 14 Gaussian mixture models (cf. Section 2.3). Near the end of the last century, a subset of eight of these models was made available as the MCLUST family, with accompanying S-PLUS software (Fraley and Raftery, 1999). The availability of this software, together with the well-known review paper of Fraley and Raftery (2002b), played an important role in popularizing model-based clustering. In fact, such was the impact of these works that the term model-based clustering became synonymous with MCLUST for several years. Another key component to the popularity of the MCLUST family is the release of an accompanying R package and, perhaps most notably, the release of `mclust` version 2 (Fraley and Raftery, 2002a).

In this chapter, the MCLUST family will be discussed within the context of the larger family described by Celeux and Govaert (1995). Elsewhere within this monograph, other families of Gaussian mixture models will be discussed, e.g, Chapters 3 and 8.

2.2   Parameter Estimation

2.2.1   Model-Based Clustering

Consider the model-based clustering paradigm, and suppose p-dimensional data vectors $\mathbf{x}_1, \ldots, \mathbf{x}_n$ are observed. Let z_{ig} denote component membership, where $z_{ig} = 1$ if observation i belongs to component g, and $z_{ig} = 0$ otherwise. Parameter estimation can be carried out using an EM algorithm, where the complete-data comprise the observed $\mathbf{x}_1, \ldots, \mathbf{x}_n$ and the labels $\mathbf{z}_1, \ldots, \mathbf{z}_n$, where $\mathbf{z}_i = (z_{i1}, \ldots, z_{iG})$. The complete-data likelihood is given by

$$\mathcal{L}_{\mathrm{c}}(\boldsymbol{\vartheta}) = \prod_{i=1}^{n} \prod_{g=1}^{G} [\pi_g \phi(\mathbf{x}_i \mid \boldsymbol{\mu}_g, \boldsymbol{\Sigma}_g)]^{z_{ig}}, \tag{2.1}$$

where $\boldsymbol{\vartheta}$ denotes the model parameters, i.e., $\boldsymbol{\vartheta} = (\boldsymbol{\pi}, \boldsymbol{\mu}_1, \ldots, \boldsymbol{\mu}_G, \boldsymbol{\Sigma}_1, \ldots, \boldsymbol{\Sigma}_G)$ with $\boldsymbol{\pi} = (\pi_1, \ldots, \pi_G)$. Taking the natural logarithm of (2.1) gives the complete-data log-likelihood

$$l_{\mathrm{c}}(\boldsymbol{\vartheta}) = \sum_{i=1}^{n} \sum_{g=1}^{G} z_{ig} \left[\log \pi_g + \log \phi(\mathbf{x}_i \mid \boldsymbol{\mu}_g, \boldsymbol{\Sigma}_g) \right]. \tag{2.2}$$

In the E-step, the expected value of the complete-data log-likelihood is updated. This amounts to replacing the z_{ig} in (2.2) by their expected values

$$\hat{z}_{ig} = \frac{\hat{\pi}_g \phi(\mathbf{x}_i \mid \hat{\boldsymbol{\mu}}_g, \hat{\boldsymbol{\Sigma}}_g)}{\sum_{h=1}^{G} \hat{\pi}_h f(\mathbf{x}_i \mid \hat{\boldsymbol{\mu}}_h, \hat{\boldsymbol{\Sigma}}_h)}, \tag{2.3}$$

for $i = 1, \ldots, n$ and $g = 1, \ldots, G$. Note that, in the E-step, we are conditioning on the current parameter estimates, hence the use of hats on the parameters in (2.3). It follows that the expected value of the complete-data log-likelihood is

$$Q(\boldsymbol{\vartheta}) = \sum_{i=1}^{n} \sum_{g=1}^{G} \hat{z}_{ig} \left[\log \pi_g - \frac{p}{2} \log 2\pi - \frac{1}{2} \log |\boldsymbol{\Sigma}_g| \right.$$

$$\left. - \frac{1}{2} \mathrm{tr} \left\{ (\mathbf{x}_i - \boldsymbol{\mu}_g)(\mathbf{x}_i - \boldsymbol{\mu}_g)' \boldsymbol{\Sigma}_g^{-1} \right\} \right]$$

$$= \sum_{g=1}^{G} n_g \log \pi_g - \frac{np}{2} \log 2\pi - \sum_{g=1}^{G} \frac{n_g}{2} \log |\boldsymbol{\Sigma}_g| - \sum_{g=1}^{G} \frac{n_g}{2} \mathrm{tr} \left\{ \mathbf{S}_g \boldsymbol{\Sigma}_g^{-1} \right\},$$

where $n_g = \sum_{i=1}^{n} \hat{z}_{ig}$ and

$$\mathbf{S}_g = \frac{1}{n_g} \sum_{i=1}^{n} \hat{z}_{ig} (\mathbf{x}_i - \boldsymbol{\mu}_g)(\mathbf{x}_i - \boldsymbol{\mu}_g)'.$$

In the M-step, the model parameters are updated. Now, maximizing $Q(\boldsymbol{\vartheta})$ with respect to π_g, $\boldsymbol{\mu}_g$, and $\boldsymbol{\Sigma}_g$ yields the updates

$$\hat{\pi}_g = \frac{n_g}{n}, \quad \hat{\boldsymbol{\mu}}_g = \frac{1}{n_g} \sum_{i=1}^{n} \hat{z}_{ig} \mathbf{x}_i, \quad \text{and} \quad \hat{\boldsymbol{\Sigma}}_g = \frac{1}{n_g} \sum_{i=1}^{n} \hat{z}_{ig} (\mathbf{x}_i - \hat{\boldsymbol{\mu}}_g)(\mathbf{x}_i - \hat{\boldsymbol{\mu}}_g)'.$$

The EM algorithm for Gaussian model-based clustering alternates between the E- and M-steps until convergence.

EM Algorithm for Gaussian Model-Based Clustering

initialize $\hat{z}_{ig}$
while convergence criterion not met
    update $\hat{\pi}_g = n_g/n$
    update $\hat{\boldsymbol{\mu}}_g = (1/n_g) \sum_{i=1}^{n} \hat{z}_{ig} \mathbf{x}_i$
    update $\hat{\boldsymbol{\Sigma}}_g = (1/n_g) \sum_{i=1}^{n} \hat{z}_{ig} (\mathbf{x}_i - \hat{\boldsymbol{\mu}}_g)(\mathbf{x}_i - \hat{\boldsymbol{\mu}}_g)'$
    update $\hat{z}_{ig}$
    check convergence criterion
end while

2.2.2   Model-Based Classification

Consider the model-based classification paradigm, and suppose p-dimensional data vectors $\mathbf{x}_1, \ldots, \mathbf{x}_n$ are observed. Suppose that k of the n observations are labelled as belonging to one of G classes and order them, without loss of generality, so that the first k are labelled. As with the labelled points, each of the unlabelled $\mathbf{x}_{k+1}, \ldots, \mathbf{x}_n$ may come from one of G classes; however, it is also possible that some of the unlabelled $\mathbf{x}_{k+1}, \ldots, \mathbf{x}_n$ come from a class, or classes, that is not represented within the labelled observations. To allow for such situations, it can be assumed that the unlabelled observations come from one of $H > G$ classes.

Parameter estimation is usually carried out using an EM algorithm, where the complete-data comprise the observed $\mathbf{x}_1, \ldots, \mathbf{x}_k, \mathbf{x}_{k+1} \ldots \mathbf{x}_n$, the known labels $\mathbf{z}_1, \ldots, \mathbf{z}_k$, and the unknown labels $\mathbf{z}_{k+1} \ldots \mathbf{z}_n$, where $\mathbf{z}_i = (z_{i1}, \ldots, z_{iH})$ and $H \geq G$. Note that, in cases where $H > G$, it is necessarily true that $z_{ih} = 0$ when $i = 1, \ldots, k$ and $h > G$. The complete-data likelihood is given by

$$\mathcal{L}_c(\boldsymbol{\vartheta}) = \prod_{i=1}^{k} \prod_{g=1}^{G} [\pi_g \phi(\mathbf{x}_i \mid \boldsymbol{\mu}_g, \boldsymbol{\Sigma}_g)]^{z_{ig}} \prod_{j=k+1}^{n} \prod_{h=1}^{H} [\pi_h \phi(\mathbf{x}_j \mid \boldsymbol{\mu}_h, \boldsymbol{\Sigma}_h)]^{z_{jh}}, \quad (2.4)$$

where $H \geq G$, and $\boldsymbol{\vartheta}$ denotes the model parameters. Taking the natural logarithm of (2.4) gives the complete-data log-likelihood

$$l_c(\vartheta) = \sum_{i=1}^{k} \sum_{g=1}^{G} z_{ig} \left[\log \pi_g + \log \phi(\mathbf{x}_i \mid \boldsymbol{\mu}_g, \boldsymbol{\Sigma}_g) \right]$$

$$+ \sum_{j=k+1}^{n} \sum_{h=1}^{H} z_{jh} \left[\log \pi_h + \log \phi(\mathbf{x}_j \mid \boldsymbol{\mu}_h, \boldsymbol{\Sigma}_h) \right]. \tag{2.5}$$

In the E-step, the expected value of the complete-data log-likelihood is updated; again, this is conditional on the current parameter estimates. This amounts to replacing the z_{jh} in (2.5) by their expected values, i.e., (2.3), for $j = k+1, \ldots, n$ and $h = 1, \ldots, H$.

In the M-step, the model parameters are updated. Now, maximizing the expected value of the complete-data log-likelihood with respect to π_g, $\boldsymbol{\mu}_g$, and $\boldsymbol{\Sigma}_g$ yields the updates

$$\hat{\pi}_g = \frac{n_g}{n}, \qquad\qquad \hat{\boldsymbol{\mu}}_g = \frac{1}{n_g} \left[\sum_{i=1}^{k} z_{ig} \mathbf{x}_i + \sum_{j=k+1}^{n} \hat{z}_{jg} \mathbf{x}_j \right],$$

and

$$\hat{\boldsymbol{\Sigma}}_g = \frac{1}{n_g} \left[\sum_{i=1}^{k} z_{ig} (\mathbf{x}_i - \hat{\boldsymbol{\mu}}_g)(\mathbf{x}_i - \hat{\boldsymbol{\mu}}_g)' + \sum_{j=k+1}^{n} \hat{z}_{jg} (\mathbf{x}_j - \hat{\boldsymbol{\mu}}_g)(\mathbf{x}_j - \hat{\boldsymbol{\mu}}_g)' \right],$$

$$\tag{2.6}$$

for $g = 1, \ldots, H$, where

$$n_g = \sum_{i=1}^{k} z_{ig} + \sum_{j=k+1}^{n} \hat{z}_{jg}.$$

The EM algorithm for Gaussian model-based classification alternates between the E- and M-steps until convergence, as outlined below.

EM Algorithm for Gaussian Model-Based Classification

initialize $\hat{z}_{jg}$ for $j > k$
while convergence criterion not met
    update $\hat{\pi}_g = n_g/n$
    update $\hat{\boldsymbol{\mu}}_g = \left[\sum_{i=1}^{k} z_{ig} \mathbf{x}_i + \sum_{j=k+1}^{n} \hat{z}_{jg} \mathbf{x}_j \right] / n_g$
    update $\hat{\boldsymbol{\Sigma}}_g$ according to (2.6)
    update $\hat{z}_{jg}$ for $j > k$
    check convergence criterion
end while

2.2.3 Model-Based Discriminant Analysis

Similar to the model-based classification paradigm (Section 2.2.2), suppose p-dimensional $\mathbf{x}_1, \ldots, \mathbf{x}_n$ are observed such that k of these n observations are labelled as belonging to one of G classes. Again, order these observations, without loss of generality, so that the first k are labelled. Within the model-based discriminant analysis paradigm, only the k labelled observations are used to estimate the model parameters and thence to estimate component memberships for the remaining $n - k$ observations. To ease the notational burden, first assume that each class is taken as corresponding to one component in the mixture model.

The likelihood for the k labelled observations is formed based on the labelled $\mathbf{x}_1, \ldots, \mathbf{x}_k$ and the associated labels $\mathbf{z}_1, \ldots, \mathbf{z}_k$:

$$\mathcal{L}(\vartheta) = \prod_{i=1}^{k} \prod_{g=1}^{G} \left[\pi_g \phi(\mathbf{x}_i \mid \boldsymbol{\mu}_g, \boldsymbol{\Sigma}_g) \right]^{z_{ig}}. \tag{2.7}$$

Maximizing the log-likelihood associated with (2.7) gives the updates

$$\hat{\pi}_g = \frac{n_g}{n}, \qquad \hat{\boldsymbol{\mu}}_g = \frac{1}{n_g} \sum_{i=1}^{k} z_{ig} \mathbf{x}_i, \qquad \hat{\boldsymbol{\Sigma}}_g = \frac{1}{n_g} \sum_{i=1}^{k} z_{ig} (\mathbf{x}_i - \hat{\boldsymbol{\mu}}_g)(\mathbf{x}_i - \hat{\boldsymbol{\mu}}_g)',$$

where $n_g = \sum_{i=1}^{k} z_{ig}$. The expected values

$$\hat{z}_{jg} := \frac{\hat{\pi}_g \phi(\mathbf{x}_j \mid \hat{\boldsymbol{\mu}}_g, \hat{\boldsymbol{\Sigma}}_g)}{\sum_{h=1}^{G} \hat{\pi}_h \phi(\mathbf{x}_j \mid \hat{\boldsymbol{\mu}}_h, \hat{\boldsymbol{\Sigma}}_h)}, \tag{2.8}$$

for $j = k + 1, \ldots, n$ and $g = 1, \ldots, G$, are then used as a discriminant rule to classify the unlabelled $\mathbf{x}_{k+1}, \ldots, \mathbf{x}_n$.

Recall that the likelihood in (2.7) assumes that each known class is associated with one component, which is effectively straightforward discriminant analysis. It is possible to relax this assumption by first performing a model-based clustering analysis for each known class (cf. Section 1.3). Specifically, a mixture model is fitted to the labelled observations in each known class and the number of components is determined via some criterion (cf. Section 2.4) so that $\mathcal{G}_g \geq 1$ components are fitted to the gth class. It is convenient to write $\mathcal{G} = \mathcal{G}_1 + \cdots + \mathcal{G}_G$ and, in this case, the likelihood is as in (2.7) but with G replaced by $\mathcal{G}$. In the case where a class may be represented by multiple components, model-based discriminant analysis proceeds as described on the next page.

As usual, the predicted classifications may be hardened to MAP classifications. Note that it will later be argued that it is often desirable to set $\mathcal{G}_g = 1$ for all $g \in \{1, \ldots, G\}$; cf., *inter alia*, Section 9.2.

Summary of Gaussian Model-Based Discriminant Analysis

for g in 1 to G

    carry out model-based clustering for $\mathbf{x}_1, \ldots, \mathbf{x}_k$ in class g

    choose a $\mathcal{G}_g$-component model, e.g., using the BIC (cf. Section 2.4)

    record corresponding labels LAB_g

end for

compute $\mathcal{G} = \mathcal{G}_1 + \cdots + \mathcal{G}_G$

fit a $\mathcal{G}$-component mixture to $\mathbf{x}_1, \ldots, \mathbf{x}_k$ using labels $\mathrm{LAB}_1, \ldots, \mathrm{LAB}_G$

record resulting parameter estimates $\hat{\boldsymbol{\pi}}$, $\hat{\boldsymbol{\mu}}_1, \ldots, \hat{\boldsymbol{\mu}}_\mathcal{G}$, $\hat{\boldsymbol{\Sigma}}_1, \ldots, \hat{\boldsymbol{\Sigma}}_\mathcal{G}$

for $j = k + 1$ to n

    for g in 1 to $\mathcal{G}$

        compute $\hat{z}_{jg} = \hat{\pi}_g \phi(\mathbf{x}_j \mid \hat{\boldsymbol{\mu}}_g, \hat{\boldsymbol{\Sigma}}_g) / \sum_{h=1}^{\mathcal{G}} \hat{\pi}_h \phi(\mathbf{x}_j \mid \hat{\boldsymbol{\mu}}_h, \hat{\boldsymbol{\Sigma}}_h)$

    end for

    assign $\mathbf{x}_j$ to the class corresponding to component $h = \arg\max_g \hat{z}_{jg}$

end for

2.2.4 Initialization via Deterministic Annealing

A common criticism of the EM algorithm for model-based applications is that the singularity-riddled likelihood surface makes parameter estimation unreliable and heavily dependent on the starting values, cf. Titterington et al. (1985). To overcome this problem, Zhou and Lange (2010) propose a deterministic annealing algorithm that makes the likelihood surface flatter by introducing an auxiliary variable $v \in (0, 1)$, which is drawn from an increasing sequence of user-specified values. The length of this sequence determines how many deterministic annealing iterations will be preformed prior to the beginning of the EM algorithm. The deterministic annealing algorithm is similar to the EM algorithm except that the quantity

$$\hat{z}_{ig}^*(v) := \frac{[\hat{\pi}_g \phi(\mathbf{x}_i \mid \hat{\boldsymbol{\mu}}_g, \hat{\boldsymbol{\Sigma}}_g)]^v}{\sum_{h=1}^{G} [\hat{\pi}_h \phi(\mathbf{x}_i \mid \hat{\boldsymbol{\mu}}_h, \hat{\boldsymbol{\Sigma}}_h)]^v} \tag{2.9}$$

is used in place of $\hat{z}_{ig}$ in each E-step. When $v = 1$, the EM algorithm is obtained and so deterministic annealing can be viewed as a progression towards the EM algorithm. The annealing algorithm can be initialized using random starting values for $\hat{\pi}_g$, $\hat{\boldsymbol{\mu}}_g$, and $\hat{\boldsymbol{\Sigma}}_g$. Deterministic annealing is illustrated in Section 2.6.

2.2.5 Stopping Rules

There are several options for stopping an EM algorithm. One popular approach is to stop an EM algorithm based on lack of progress in the log-likelihood, i.e., stopping the algorithm when

$$l^{(k+1)} - l^{(k)} < \epsilon, \tag{2.10}$$

for ϵ small, where $l^{(k)}$ is the (observed) log-likelihood value from iteration k. This stopping rule can work very well when the log-likelihood increases and then plateaus at the maximum likelihood estimate (e.g., Figure 2.1).

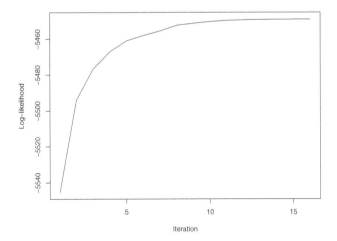

Figure 2.1
Stereotypical plot of log-likelihood value versus iteration number for an EM algorithm.

However, likelihood values do not necessarily progress, iteration-on-iteration, as nicely as depicted in Figure 2.1. For this reason, it can be worth considering alternatives. Böhning et al. (1994), Lindsay (1995), and McNicholas et al. (2010) consider convergence criteria based on Aitken's acceleration (Aitken, 1926). Aitken's acceleration at iteration k is

$$a^{(k)} = \frac{l^{(k+1)} - l^{(k)}}{l^{(k)} - l^{(k-1)}}, \tag{2.11}$$

and an asymptotic estimate of the log-likelihood at iteration $k + 1$ can be computed via

$$l_\infty^{(k+1)} = l^{(k)} + \frac{l^{(k+1)} - l^{(k)}}{1 - a^{(k)}}. \tag{2.12}$$

Note that "asymptotic" here refers to the iteration number so that (2.12) can be interpreted as an estimate, at iteration $k + 1$, of the ultimate value of the log-likelihood. The algorithm can be considered to have converged when

$$|l_\infty^{(k+1)} - l_\infty^{(k)}| < \epsilon \tag{2.13}$$

(Böhning et al., 1994), or when

$$l_\infty^{(k)} - l^{(k)} < \epsilon \tag{2.14}$$

(Lindsay, 1995), or when

$$l_\infty^{(k+1)} - l^{(k)} < \epsilon \qquad (2.15)$$

provided that this difference is positive (McNicholas et al., 2010). Note that the caveat regarding the positivity of the difference in (2.15) is important but sometimes not stated explicitly. The reason it is important is that it would not make sense to stop the algorithm when $a^{(k)} > 1$ and this is the only circumstance under which the difference in (2.15) can be negative — a fact that follows from rearrangement of (2.12). For this reason, the stopping rule

$$|l_\infty^{(k+1)} - l^{(k)}| < \epsilon$$

should not be used.

In general, the criteria in (2.13), (2.14), and (2.15) will give the same or very similar results. McNicholas et al. (2010) prefer the criterion in (2.15) because it is easy to see that it is necessarily at least as strict as lack of progress, i.e., (2.10), in the neighbourhood of a maximum. In many cases, e.g., Figure 2.1, the criteria in (2.13), (2.14), and (2.15) will essentially return the same result as lack of progress; however, they may lead to the algorithm running longer than really necessary. For this reason, and because they require additional computation, some argue against their use.

Of course, if one knew *a priori* that the situation would be as depicted in Figure 2.1, then lack of progress would be preferable. However, it is possible that the likelihood will "jump" once or more, as depicted in Figure 2.2. In these cases, there is the possibility that using lack of progress in the likelihood will lead to the algorithm stopping prematurely. A simple analogy as to why the criteria in (2.14) and (2.15) handle likelihood progressions with steps, e.g., Figure 2.2, is that they are forward looking, i.e., they compare the current log-likelihood value to an estimate of the ultimate value. Lack of progress, on the other hand, is backward looking, i.e., it compares the current and previous log-likelihood values.

2.3 Gaussian Parsimonious Clustering Models

A p-dimensional random variable following a G-component Gaussian mixture model has a total of

$$G - 1 + Gp + \frac{Gp(p+1)}{2}$$

free parameters: $G - 1$ from the mixing proportions, Gp from the means, and $Gp(p+1)/2$ from the covariance matrices. Because there are $p(p+1)/2$ free parameters for each Σ_g, it is a natural focus for the introduction of parsimony. One could consider the Gaussian mixture model with $\Sigma_g = \lambda_g \mathbf{I}_p$, $\Sigma_g = \lambda \mathbf{I}_p$, or $\Sigma_g = \Sigma$ (cf. Section 1.2). Other cluster constraints are discussed by Banfield

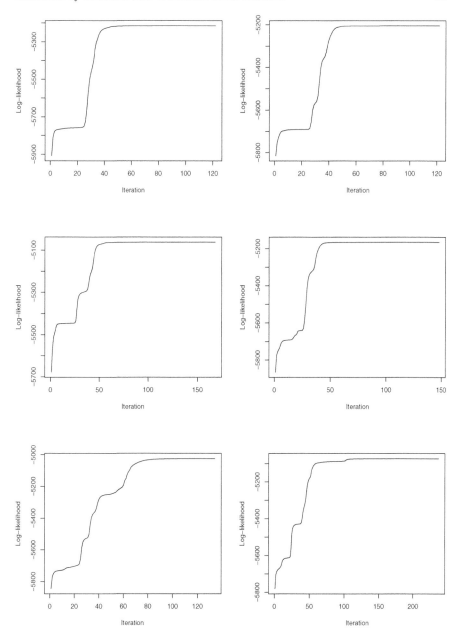

Figure 2.2
Plots of log-likelihood value versus iteration, with varying numbers of steps, for EM algorithms.

and Raftery (1993), who consider an eigen-decomposition of the component covariance matrices, i.e.,

$$\boldsymbol{\Sigma}_g = \lambda_g \boldsymbol{\Gamma}_g \boldsymbol{\Delta}_g \boldsymbol{\Gamma}'_g, \tag{2.16}$$

where $\lambda_g = |\boldsymbol{\Sigma}_g|^{1/p}$, $\boldsymbol{\Gamma}_g$ is the matrix of eigenvectors of $\boldsymbol{\Sigma}_g$, and $\boldsymbol{\Delta}_g$ is a diagonal matrix, such that $|\boldsymbol{\Delta}_g| = 1$, containing the normalized eigenvalues of $\boldsymbol{\Sigma}_g$ in decreasing order. Note that the columns of $\boldsymbol{\Gamma}_g$ are ordered to correspond to the elements of $\boldsymbol{\Delta}_g$. As Banfield and Raftery (1993) point out, the constituent elements of the decomposition in (2.16) can be viewed in the context of the geometry of the component, where λ_g represents the volume in p-space, $\boldsymbol{\Delta}_g$ the shape, and $\boldsymbol{\Gamma}_g$ the orientation. By imposing constraints on the elements of the decomposed covariance structure in (2.16), Celeux and Govaert (1995) introduce a family of Gaussian parsimonious clustering models (GPCMs; Table 2.1).

Table 2.1

The type, nomenclature, and covariance structure for each member of the GPCM family.

Type	Model	Volume	Shape	Orientation	$\boldsymbol{\Sigma}_g$
Spherical	EII	Equal	Spherical		$\lambda \mathbf{I}$
	VII	Variable	Spherical		$\lambda_g \mathbf{I}$
Diagonal	EEI	Equal	Equal	Axis-Aligned	$\lambda \boldsymbol{\Delta}$
	VEI	Variable	Equal	Axis-Aligned	$\lambda_g \boldsymbol{\Delta}$
	EVI	Equal	Variable	Axis-Aligned	$\lambda \boldsymbol{\Delta}_g$
	VVI	Variable	Variable	Axis-Aligned	$\lambda_g \boldsymbol{\Delta}_g$
General	EEE	Equal	Equal	Equal	$\lambda \boldsymbol{\Gamma} \boldsymbol{\Delta} \boldsymbol{\Gamma}'$
	VEE	Variable	Equal	Equal	$\lambda_g \boldsymbol{\Gamma} \boldsymbol{\Delta} \boldsymbol{\Gamma}'$
	EVE	Equal	Variable	Equal	$\lambda \boldsymbol{\Gamma} \boldsymbol{\Delta}_g \boldsymbol{\Gamma}'$
	EEV	Equal	Equal	Variable	$\lambda \boldsymbol{\Gamma}_g \boldsymbol{\Delta} \boldsymbol{\Gamma}'_g$
	VVE	Variable	Variable	Equal	$\lambda_g \boldsymbol{\Gamma} \boldsymbol{\Delta}_g \boldsymbol{\Gamma}'$
	VEV	Variable	Equal	Variable	$\lambda_g \boldsymbol{\Gamma}_g \boldsymbol{\Delta} \boldsymbol{\Gamma}'_g$
	EVV	Equal	Variable	Variable	$\lambda \boldsymbol{\Gamma}_g \boldsymbol{\Delta}_g \boldsymbol{\Gamma}'_g$
	VVV	Variable	Variable	Variable	$\lambda_g \boldsymbol{\Gamma}_g \boldsymbol{\Delta}_g \boldsymbol{\Gamma}'_g$

The 14 GPCM models can be thought of as belonging to one of three categories: spherical, diagonal, and general. Of these three categories, only the eight general models have flexibility in their orientation, i.e., do not assume that the variables are independent. In terms of parsimony, the spherical and diagonal models — there are six in all — have $\mathcal{O}(p)$ free covariance parameters; however, the eight general models have $\mathcal{O}(p^2)$ covariance parameters (Table 2.2). From this observation, one view of the GPCM family of models emerges, i.e., six parsimonious models that assume the variables are independent and eight models that do not. Whether the eight models that do not

assume independence are parsimonious is debatable. Certainly, seven of them have fewer parameters than the full model (VVV; Table 2.1); however, the fact that there are $\mathcal{O}(p^2)$ free covariance parameters in each of these models means it is debatable that they are truly parsimonious. The following is probably a fair position concerning the eight general models: seven of them are more parsimonious than the VVV model.

Table 2.2
The number of free covariance parameters for each member of the GPCM family.

Model	Σ_g	Number of Free Covariance Parameters
EII	λI	1
VII	$\lambda_g I$	G
EEI	$\lambda \Delta$	p
VEI	$\lambda_g \Delta$	$G + p - 1$
EVI	$\lambda \Delta_g$	$1 + G(p-1)$
VVI	$\lambda_g \Delta_g$	Gp
EEE	$\lambda \Gamma \Delta \Gamma'$	$p(p+1)/2$
VEE	$\lambda_g \Gamma \Delta \Gamma'$	$G + p - 1 + p(p-1)/2$
EVE	$\lambda \Gamma \Delta_g \Gamma'$	$1 + G(p-1) + p(p-1)/2$
EEV	$\lambda \Gamma_g \Delta \Gamma'_g$	$p + Gp(p-1)/2$
VVE	$\lambda_g \Gamma \Delta_g \Gamma'$	$Gp + p(p-1)/2$
VEV	$\lambda_g \Gamma_g \Delta \Gamma'_g$	$G + p - 1 + Gp(p-1)/2$
EVV	$\lambda \Gamma_g \Delta_g \Gamma'_g$	$1 + G(p-1) + Gp(p-1)/2$
VVV	$\lambda_g \Gamma_g \Delta_g \Gamma'_g$	$Gp(p+1)/2$

For most members of the GPCM family, parameter estimation can be carried out via an EM algorithm as described for the VVV model in Section 2.2.1. The estimation of the covariance parameters, i.e., λ_g, Γ_g, and Δ_g, within the EM framework is outlined by Celeux and Govaert (1995). Browne and McNicholas (2014c) point out that the algorithms Celeux and Govaert (1995) use for the EVE and VVE models are computationally infeasible in higher dimensions. They develop alternative algorithms for these models, based on an accelerated line search on the orthogonal Stiefel manifold (see Browne and McNicholas, 2014c, for details). Browne and McNicholas (2014a) develop another approach, using fast majorization-minimization algorithms, for the EVE and VVE models and it is this approach that is implemented in the `mixture` package for R. Details on this latter approach are given in Browne and McNicholas (2014a). See Section 9.8 for some further discussion on alternative approaches, i.e., beyond the EM algorithm, to parameter estimation for the GPCM family.

2.4   Model Selection

A typical application of the GPCM family of models consists of running each of the models (Table 2.1) for a range of values of G. Then the best of these models is selected using some criterion and the associated classifications are reported. The most popular criterion for this purpose is the Bayesian information criterion (BIC; Schwarz, 1978), i.e.,

$$\text{BIC} = 2l(\hat{\boldsymbol{\vartheta}}) - \rho \log n, \tag{2.17}$$

where $\hat{\boldsymbol{\vartheta}}$ is the maximum likelihood estimate of $\boldsymbol{\vartheta}$, $l(\hat{\boldsymbol{\vartheta}})$ is the maximized log-likelihood, and ρ is the number of free parameters in the model. Leroux (1992) and Keribin (2000) give theoretical results that, under certain regularity conditions, support the use of the BIC for choosing the number of components in a mixture model.

Campbell et al. (1997) use the BIC in a model-based clustering setting where $G - 1$ of the G components are Gaussian and the other one is Poisson. Dasgupta and Raftery (1998) discuss the BIC in the context of selecting the number of components in a Gaussian mixture model. Its application to this problem, i.e., selection of G, is part of the reason why the BIC has become so popular for mixture model selection in general. Dasgupta and Raftery (1998) use the BIC as a basis for an approximation to Bayes factors (Kass and Raftery, 1995), pointing out that

$$2 \log p(\mathbf{x} \mid \boldsymbol{\vartheta}_G) \approx \text{BIC}_G,$$

where $p(\mathbf{x} \mid \boldsymbol{\vartheta}_G)$ is the integrated likelihood of the data under a G-component mixture model and BIC_G is the quantity given in (2.17) with the subscript being used to emphasize that it is based on a G-component mixture. If each choice for G is *a priori* equally likely, then

$$p(\mathbf{x} \mid \boldsymbol{\vartheta}_G) \propto h(\boldsymbol{\vartheta}_G \mid \mathbf{x}),$$

where $h(\boldsymbol{\vartheta}_G \mid \mathbf{x})$ is the posterior probability of the G-component model. Dasgupta and Raftery (1998) suggest that, when comparing models, BIC differences of more than 10 constitute "very strong evidence".

The BIC has become very popular for selecting the number of components as well as the model (i.e., covariance or scale decomposition) and the number of latent factors, where relevant, in model-based clustering applications. Some recent examples of its use include work by Bouveyron et al. (2007a), McNicholas and Murphy (2008), Andrews and McNicholas (2011a), and Vrbik and McNicholas (2014). Of course, despite its popularity, the model selected by the BIC does not necessarily give the best classification performance from among the candidate models. To this end, alternatives such as the integrated completed likelihood (ICL; Biernacki et al., 2000) have been considered. The ICL can be calculated via

$$\text{ICL} \approx \text{BIC} + 2 \sum_{i=1}^{n} \sum_{g=1}^{G} \text{MAP}\{\hat{z}_{ig}\} \log \hat{z}_{ig}, \tag{2.18}$$

where

$$2 \sum_{i=1}^{n} \sum_{g=1}^{G} \text{MAP}\{\hat{z}_{ig}\} \log \hat{z}_{ig}$$

is an entropy penalty that reflects the uncertainty in the classification of observations into components.

Despite the ongoing search for alternatives to the BIC, and some dedicated comparative studies (e.g., Steele and Raftery, 2010), the BIC remains the model selection criterion of choice in most model-based clustering applications.

2.5 Merging Gaussian Components

In model-based discriminant analysis applications, the option of using more than one component to represent a class is often exercised. However, clusters are generally taken as synonymous with mixture components and a similar one-to-one relationship is commonly assumed in model-based classification. A cluster might itself be better modelled by a mixture of Gaussian distributions rather than a single Gaussian distribution and, in such cases, mixture components can be *a posteriori* merged to obtain clusters. Several different approaches have been considered for *a posteriori* merging. Hennig (2010) suggests a hierarchical merging approach, where some aggregation criterion is used. Such criteria are based on merging components to produce a unimodal distribution or to minimize misclassification probabilities. Wei and McNicholas (2015) give an approach for model averaging in model-based clustering and, in doing so, introduce a merging procedure based on maximizing the ARI with respect to a reference model (cf. Section 9.3). Baudry et al. (2010) propose a method for combining mixture components to represent one cluster. This method first fits a mixture of Gaussian distributions with G components and then successively merges mixture components together, resulting in a solution with $K \leq G$ components. The `mclust` package contains a function called `clustCombi` that implements the methodology proposed in Baudry et al. (2010). The `clustCombi` function starts from the original G-component solution (as chosen by the BIC) and merges two components according to an entropy criterion to obtain a $(G-1)$-component solution. This procedure can be repeated until the one-component solution is obtained. As suggested by Baudry et al. (2010), components can be merged until the number of clusters is equal to that obtained by the ICL, or the number of components can be selected to correspond to an "elbow" in a plot of the entropy versus G. Another merging approach is what Vrbik and McNicholas (2014) call "merging by hand", which boils down to merging components to give the best classification results; of course, this approach assumes knowledge of the true classes.

Consider the simulated bivariate data in Figure 2.3. How many clusters are there? The most sensible answer is three: two elliptical clusters centred around $(3, 3)$ and a skewed cluster with mode around $(9, 9)$. Not only is this answer the most sensible by inspection of Figure 2.3, it also corresponds to how the data were generated. In higher dimensions, it is not feasible to look at a plot and make a determination about the number of clusters; accordingly, approaches that do not depend on visual inspection are important.

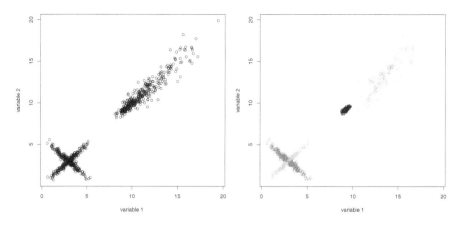

Figure 2.3
Scatter plots for data simulated from two Gaussian distributions and one skewed distribution where, on the right-hand side, shading and plotting symbols represent the predicted classifications from the fitted VVV model.

Gaussian mixture models are fitted to these data for $G = 1, \ldots, 6$ components, using `mixture` with only the VVV model. The BIC selects a $G = 5$ component mixture and the associated MAP classifications are shown in Figure 2.3. Consider the predicted classifications associated with the three Gaussian components that have been fitted to the skewed cluster, i.e., the cluster with mode around $(9, 9)$, in Figure 2.3. This is a good example of how Gaussian mixture models can be used to effectively model a skewed cluster because *a posteriori* merging of these Gaussian components will return a sensible result — some further discussion on using a Gaussian mixture to model a skewed cluster is given in Section 6.3.3. On the other hand, merging the two Gaussian components centred around $(3, 3)$ will lead to a cluster that will, in general, not be sensible. The reason is that the presence of some overlap is usually not sufficient justification for merging a component where the two variables are strongly positively correlated with a component where they are strongly negatively correlated. In summary, consider the two graphs in Figure 2.4. Both are the result of fitting a Gaussian mixture model followed by merging, but the one on the left-hand side is clearly the more sensible solution.

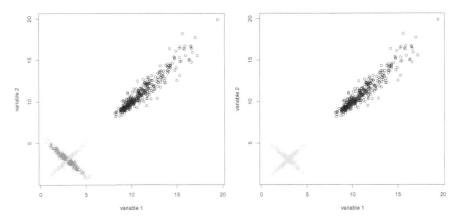

Figure 2.4
Two possible merging outcomes following the model-based clustering analysis of data simulated from two Gaussian components and one skewed component.

2.6 Illustrations

2.6.1   x2 Data

First, consider the x2 dataset from the `mixture` package. This is a two-dimensional dataset generated from an EVE model with three components, $\pi_1 = \pi_2 = \pi_3 = 1/3$, and $n = 300$ (Figure 2.5). The x2 data represent a very easy classification example and quite a straightforward clustering example. They are used by Browne and McNicholas (2014a) to illustrate their majorization-minimization algorithms for clustering, and especially that for the EVE model. Here the x2 data will serve as the first of several illustrations of the importance of starting values for model-based clustering applications — even straightforward examples. All 14 GPCMs are fitted to these data using the `gpcm()` function from `mixture`, for $G = 1, \ldots, 5$, using k-means starting values; the reader is referred to Steinley (2006) for a detailed review of work on k-means clustering (MacQueen, 1967; Hartigan and Wong, 1979). The BIC selects a $G = 3$ component EVE model with MAP classifications corresponding to the true model, save for the single point that one would expect to be misclassified (cf. Figure 2.5). The BIC for the selected model is -1981.47.

Repeating this process with five random starting values leads to the same model.

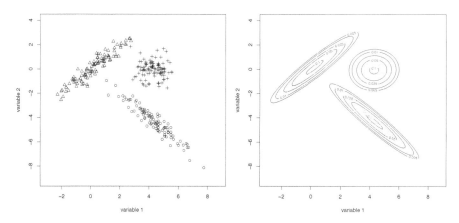

Figure 2.5
Scatter plots depicting the x2 dataset, where plotting symbols reflect the components used to generate the data (left), and with contours from the fitted GPCM model where plotting symbols reflect predicted classifications (right).

Looking at the right-hand side of Figure 2.5 raises an interesting question about the EVE model and the interpretation of the GPCM models in general. What does it mean for the orientation of the components to be equal (i.e., $\mathbf{\Gamma}_g = \mathbf{\Gamma}$)? Figure 2.5 tells us that the answer is not that the components must literally have the same orientation, i.e., the components need not be parallel in their orientation. Considering the EVI (i.e., $\mathbf{\Gamma}_g = \mathbf{I}$) model is instructive; here components must be axis-aligned, e.g., one component might align with the x-axis and another with the y-axis. One interpretation of the EVE model, or any other model with $\mathbf{\Gamma}_g = \mathbf{\Gamma}$, is that these axes can be rotated. Put differently, when $\mathbf{\Gamma}_g = \mathbf{\Gamma}$, the components must vary about the same (principal) axes, as illustrated in Figure 2.6, but they need not be parallel in their orientation. Of course, the same argument holds in higher dimensions.

Now consider deterministic annealing starts for these data. The gpcm() function allows specification of an annealing vector, i.e., a vector of values for v in (2.9), using the argument start. The results (Table 2.3) suggest that it is somewhat preferable here to only use values of v nearer to one. Note that all but the $G = 4$ component VVV model (ARI = 0.97) give the same classification results as the model with k-means starts.

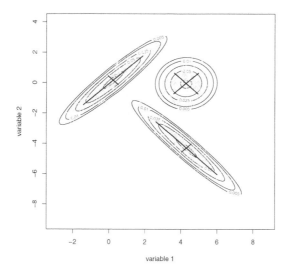

variable 1

Figure 2.6
Scatter plot depicting the x2 dataset with contours and orientation axes from
the selected three-component EVE model, where plotting symbols reflect the
associated MAP classifications.

Table 2.3
Results from applying the GPCM models to the x2 data using deterministic
annealing starts, via the gpcm() function, for different values of start.

start	Selected Model	BIC
c(0.1,0.2,0.3,0.4,0.5,0.7,0.9,1)	VVV, $G = 4$	−2034.20
c(0.1,0.2,0.5,0.9,1)	EVV, $G = 3$	−1992.48
c(0.2,0.5,0.9,1)	EVE, $G = 3$	−1981.47
c(0.5,0.9,1)	EVE, $G = 3$	−1981.47
c(0.9,1)	EVE, $G = 3$	−1981.47

2.6.2   Banknote Data

Next, consider the banknote dataset from the mclust package. The dataset
contains six measurements, all in mm, made on 100 genuine and 100 counterfeit Swiss 1000-franc banknotes (Figure 2.7). The discrimination line between
Top and Diagonal (cf. Figure 2.7) makes this a straightforward classification
example. However, it is more challenging as a clustering example; accordingly,
the banknote data will serve as another illustration of the importance of starting values for model-based clustering applications. All 14 GPCMs are fitted
to these data using the gpcm() function from mixture, for $G = 1, \ldots, 5$, using
k-means starting values. The BIC selects a $G = 3$ component EVE model with

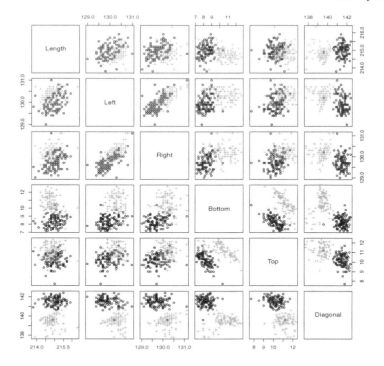

Figure 2.7
Pairs plot depicting the `banknote` data, where plotting symbol and shading reflect class, i.e., genuine or counterfeit.

BIC = −1608.45. A cross-tabulation of the corresponding MAP classifications versus the true classes is given in Table 2.4. Using five random starts, the BIC selects a $G = 4$ component EEE model with BIC = −1607.31 (Table 2.5).

Consider deterministic annealing starts for these data, using the `start` argument in the `gpcm()` function. The same annealing vectors are used as for the `x2` data (cf. Table 2.3). In all five cases, the same model is selected as with k-means starts, i.e., an EVE model with $G = 3$ components and the MAP classifications in Table 2.4.

Table 2.4
Cross-tabulation of the MAP classifications (A–C) associated with the selected GPCM, using k-means starts, against true classes for the `banknote` data.

	A	B	C
Counterfeit	85	15	0
Genuine	0	1	99

Table 2.5
Cross-tabulation of the MAP classifications (A–D) associated with the selected GPCM, using five random starts, against true classes for the `banknote` data.

	A	B	C	D
Counterfeit	85	15	0	0
Genuine	0	1	24	75

2.6.3 Female Voles Data

The female voles data are available as `f.voles` within the `Flury` package. These data contain age and six morphometric measurements for 86 female voles from two species: *Microtus californicus* and *M. ochrogaster* (Figure 2.8). The data are scaled prior to analysis.

The discrimination line between B4.Interorbital and H1.Skull (cf. Figure 2.8) makes this a straightforward classification example. However, it is more challenging as a clustering example; accordingly, the `f.voles` data will serve as a final illustration of GPCM model-based clustering. All 14 GPCMs are fitted to these data using the `gpcm()` function from `mixture`, for $G = 1, \ldots, 5$, using k-means starting values. The BIC selects a $G = 2$ component EEE model with BIC $= -1316.70$. A cross-tabulation of the corresponding MAP classifications versus the true classes is given in Table 2.6. Using five random starts leads to a different result; the BIC selects a $G = 2$ component EVE model with BIC $= -1321.72$. A cross-tabulation of the corresponding MAP classifications versus the true classes is given in Table 2.7. Although the model using random starts gives slightly better classification performance, the BIC is better for the model with k-means starts.

Table 2.6
Cross-tabulation of the MAP classifications (A, B) associated with the selected GPCM, using k-means starts, against true classes for the `f.voles` data.

	A	B
M. californicus	41	0
M. ochrogaster	2	43

Consider deterministic annealing starts for these data, using the `start` argument in the `gpcm()` function with the same annealing vectors as used for the `x2` and `banknote` datasets. This time, the results (Table 2.8) suggest that it is somewhat preferable here to include values of v nearer zero. Note that all of the EVE models with $G = 2$ components give the same classification results as in Table 2.7. Note also that, although two annealing vectors returned

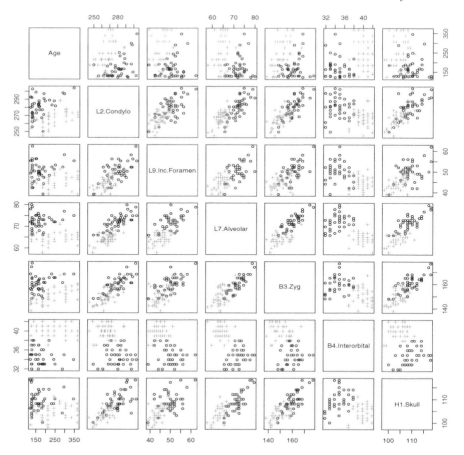

Figure 2.8
Scatter plot for the `f.voles` data, where plotting symbol and shading reflect species.

Table 2.7
Cross-tabulation of the MAP classifications (A, B) associated with the selected GPCM, using random starts, against true classes for the `f.voles` data.

	A	B
M. californicus	41	0
M. ochrogaster	1	44

models with $G = 1$ component, these models would not be selected based on their BIC values compared to other models fitted.

Table 2.8
Results from applying the GPCM models to the `f.voles` data using deterministic annealing starts, via the `gpcm()` function, for different values of `start`.

start	Selected Model	BIC
c(0.1,0.2,0.3,0.4,0.5,0.7,0.9,1)	EVE, $G = 2$	-1321.72
c(0.1,0.2,0.5,0.9,1)	EVE, $G = 2$	-1321.72
c(0.2,0.5,0.9,1)	EVE, $G = 2$	-1321.72
c(0.5,0.9,1)	$G = 1$	-1347.36
c(0.9,1)	$G = 1$	-1347.36

2.6.4 Italian Olive Oil Data

Forina and Tiscornia (1982) and Forina et al. (1983) report the percentage composition of eight fatty acids found by lipid fraction of 572 Italian olive oils (Table 2.9).

Table 2.9
Eight fatty acids extracted from the Italian olive oils.

Palmitic acid	Palmitoleic acid	Stearic acid
Oleic acid	Linoleic acid	Linolenic acid
Arachidic acid	Eicosenoic acid	

Broadly, the data come from three regions, Southern Italy, Sardinia, and Northern Italy, but within these regions there are a number of different areas. Within Southern Italy are North Apulia, Calabria, South Apulia, and Sicily; Sardinia is broken into Inland Sardinia and Coastal Sardinia; and Northern Italy comprises Umbria, East Liguria, and West Liguria. These data are available within the pgmm package, and they have previously been used to illustrate classification approaches (e.g., McNicholas, 2010).

McNicholas (2010) explains that the problem of classifying these olive oils into the appropriate area, or region, is an important one in food authenticity studies because oils from some areas have more commercial value than others. As mentioned by Cook and Swayne (2007), the classification into the nine areas is much more difficult than the more coarse classification into the three regions; a pairs plot of the data with the regions highlighted is given in Figure 2.9. Here the focus is on the more difficult problem of classification by area.

A classification scenario is simulated by taking every fourth olive oil to be unlabelled, with the effect that precisely 25% of the oils are unlabelled. All GPCM models are run for $G = 9$ and the BIC (-5036.71) selects a VVE model. The classification performance of this model is very good (Table 2.10; ARI $= 0.96$). Although only four oils are misclassified, one of these is mis-

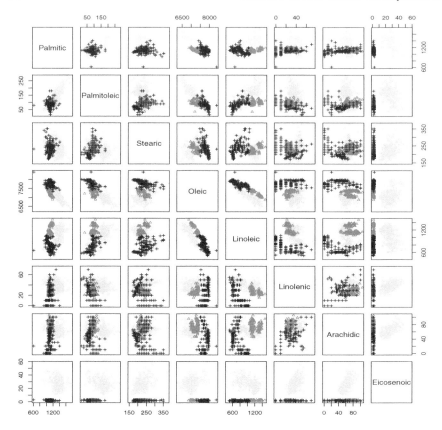

Figure 2.9
Pairs plot depicting the `olive` data, where plotting symbol and shading reflect region.

classified out-of-region; i.e., if one considers the 4-2-3 block-diagonal structure of Table 2.10, it becomes apparent that one oil from East Liguria, in Northern Italy, is classified into the same group as the oils from North Apulia, in Southern Italy.

Using the same data, i.e., the `olive` data with every fourth olive oil taken to be unlabelled, model-based discriminant analysis is carried out using the `MclustDA()` function in `mclust`. The classification performance of this model is not as good as for model-based classification (Table 2.11; ARI = 0.68). Furthermore, 10 of the 34 misclassified oils are misclassified out-of-region, including nine oils from Coastal Sardinia that are put in the same class as oils from Southern Italy. Details on the model selected by `MclustDA()` are given in Table 2.12, from which it is clear that classes fitted using several components produce many more misclassifications. From the four areas modelled using $G = 1$ component each, there is a total of only one misclassified oil. One

Table 2.10
Cross-tabulation of the MAP classifications (A–I) associated with the selected GPCM model against true classes for the `olive` data, where every fourth oil is treated as unlabelled.

	A	B	C	D	E	F	G	H	I
North Apulia	6	0	0	0	0	0	0	0	0
Calabria	0	14	0	0	0	0	0	0	0
South Apulia	0	0	50	1	0	0	0	0	0
Sicily	0	1	0	8	0	0	0	0	0
Inland Sardinia	0	0	0	0	16	0	0	0	0
Coastal Sardinia	0	0	0	0	0	9	0	0	0
East Liguria	1	0	0	0	0	0	11	0	0
West Liguria	0	0	0	0	0	0	1	12	0
Umbria	0	0	0	0	0	0	0	0	13

area is modelled using $G = 2$ components and there are no misclassifications. The remaining four areas are modelled using $G = 4$ or $G = 5$ components each, resulting in a total of 33 misclassified oils. These results are all the more disappointing when one considers that the classification performance on the training set is excellent (Table 2.13).

Table 2.11
Cross-tabulation of the MAP classifications (A–I) associated with the fitted `MclustDA` model against true classes for the `olive` data (i.e., the test set), where every fourth oil is treated as unlabelled.

	A	B	C	D	E	F	G	H	I
North Apulia	5	0	1	0	0	0	0	0	0
Calabria	0	14	0	0	0	0	0	0	0
South Apulia	0	0	51	0	0	0	0	0	0
Sicily	2	2	3	2	0	0	0	0	0
Inland Sardinia	0	0	0	0	16	0	0	0	0
Coastal Sardinia	0	0	9	0	0	0	0	0	0
East Liguria	0	0	0	0	0	0	12	0	0
West Liguria	0	0	0	0	0	0	7	6	0
Umbria	0	1	0	0	0	0	9	0	3

Table 2.12

Details of the model fitted by `MclustDA()` for model-based discriminant analysis of the `olive` data. Note that XXX refers to an ellipsoidal multivariate Gaussian distribution and is used for $G = 1$ component models.

Classes	n_g (Training)	Model	G	Misclassified
North Apulia	19	XXX	1	1
Calabria	42	XXX	1	0
South Apulia	155	XXX	1	0
Sicily	27	VEV	4	7
Inland Sardinia	49	EEI	2	0
Coastal Sardinia	24	VEV	4	9
East Liguria	38	XXX	1	0
West Liguria	37	EEV	5	7
Umbria	38	EEV	5	10

Table 2.13

Cross-tabulation of the classifications (A–I) associated with the `MclustDA` model fitted to the training data against true classes for the `olive` data, where every fourth oil is treated as unlabelled.

	A	B	C	D	E	F	G	H	I
North Apulia	19	0	0	0	0	0	0	0	0
Calabria	0	41	1	0	0	0	0	0	0
South Apulia	0	1	154	0	0	0	0	0	0
Sicily	0	0	0	27	0	0	0	0	0
Inland Sardinia	0	0	0	0	49	0	0	0	0
Coastal Sardinia	0	0	0	0	0	24	0	0	0
East Liguria	0	0	0	0	0	0	38	0	0
West Liguria	0	0	0	0	0	0	0	37	0
Umbria	0	0	0	0	0	0	0	0	38

2.7   Comments

The Gaussian mixture model can be used to effectively cluster and classify data. This has been illustrated in the analyses in Section 2.6. The dependence on starting values has also been demonstrated and, although classification performance varies depending on starting values, the model with the best BIC value was always associated with a very good, and sometimes the best, classification performance. When clustering the x2 data, deterministic annealing starts worked a little better when the annealing vector contained larger values for v; however, they worked better for the f.voles data when smaller values were included. This said, the BIC tended to choose the best of the solutions when they varied for different deterministic annealing starts.

Other approaches besides the EM algorithm could be used for parameter estimation. These other approaches, including MM algorithms and variational approximations, can either complement or replace the EM algorithm. Some of these alternatives are discussed in Section 9.8. Although alternatives exist, it is worth noting that the EM algorithm and extensions thereof remain by far the most popular approaches for parameter estimation in model-based approaches. EM algorithms are attractive for many reasons, including ease of debugging — inserting extra E-steps is a very effective way to narrow down errors in implementations of EM algorithms. Of course, this is effective because the likelihood is non-decreasing. A deterministic annealing strategy for starting the EM algorithm was discussed in Section 2.2.4; however, many other strategies are available (see Biernacki et al., 2003, for examples).

The olive oil example (Section 2.6.4) highlighted the inherent danger in model-based discriminant analysis approaches where multiple components are used to represent one class. Specifically, notable classification error can arise, e.g., $G = 4$ components were used for the 24 oils from Coastal Sardinia in the training set, with perfect associated classifications, and yet all nine oils in the test set were misclassified. Notably, model-based classification gives excellent classification performance for the same data. Experience suggests that great care is needed when using multiple components per class; while it can work very well if the classification example is easy, e.g., the **banknote** data, or the proportion of labelled points is very high, the problems observed with the olive oil data can arise. Examples will later be given where only one (non-Gaussian) component is used for each class and discriminant analysis is very effective, e.g., the olive oil data are revisited in Section 7.7.3.

From a foundational point of view, a sort of argument in favour of mixture models for clustering was given in Section 2.1. However, it is better to view this material as the beginning of an argument as to why a cluster should be defined in terms of a component in a mixture model. This argument is elaborated upon and discussed further in Sections 9.1 and 9.2. Finally, it is interesting to note that the x2 data represent an example of data simulated from a mixture model but where the "correct" classifications cannot reasonably be taken to correspond perfectly to the model used to generate the data (cf. Figure 2.5).

3

Mixtures of Factor Analyzers and Extensions

3.1 Factor Analysis

3.1.1 The Model

Factor analysis is a data reduction technique that replaces p observed variables by $q < p$ latent factors. The method works well when the q latent factors explain a satisfactory amount of the variability in the p observed variables. In some situations, we might even have substantial dimension reduction, i.e., $q \ll p$. Factor analysis originated in the psychology literature (Spearman, 1904, 1927) and was laid out in statistical terms by Bartlett (1953) and Lawley and Maxwell (1962). Consider independent p-dimensional random variables $\mathbf{X}_1, \ldots, \mathbf{X}_n$. The factor analysis model can be written

$$\mathbf{X}_i = \boldsymbol{\mu} + \boldsymbol{\Lambda} \mathbf{U}_i + \boldsymbol{\varepsilon}_i, \tag{3.1}$$

for $i = 1, \ldots, n$, where $\boldsymbol{\Lambda}$ is a $p \times q$ matrix of factor loadings, the latent factor $\mathbf{U}_i \sim \mathrm{N}(\mathbf{0}, \mathbf{I}_q)$, and $\boldsymbol{\varepsilon}_i \sim \mathrm{N}(\mathbf{0}, \boldsymbol{\Psi})$, where $\boldsymbol{\Psi} = \mathrm{diag}(\psi_1, \psi_2, \ldots, \psi_p)$. Note that the $\mathbf{U}_i$ are independently distributed and independent of the $\boldsymbol{\varepsilon}_i$, which are also independently distributed. From (3.1), it follows that the marginal distribution of $\mathbf{X}_i$ under the factor analysis model is $\mathrm{N}(\boldsymbol{\mu}, \boldsymbol{\Lambda}\boldsymbol{\Lambda}' + \boldsymbol{\Psi})$. There are

$$pq + p - \frac{1}{2}q(q-1)$$

free parameters in the covariance matrix $\boldsymbol{\Lambda}\boldsymbol{\Lambda}' + \boldsymbol{\Psi}$ (Lawley and Maxwell, 1962). Therefore, the reduction in free covariance parameters under the factor analysis model is

$$\frac{1}{2}p(p+1) - \left[pq + p - \frac{1}{2}q(q-1) \right] = \frac{1}{2}\left[(p-q)^2 - (p+q) \right], \tag{3.2}$$

and there is a reduction in the number of free parameters provided that (3.2) is positive, i.e., provided that

$$(p-q)^2 > (p+q).$$

The log-likelihood for $\mathbf{x}_1, \mathbf{x}_2, \ldots, \mathbf{x}_n$ from a factor analysis model is

$$l(\boldsymbol{\mu}, \boldsymbol{\Lambda}, \boldsymbol{\Psi}) = \sum_{i=1}^{n} \log \phi(\mathbf{x}_i \mid \boldsymbol{\mu}, \boldsymbol{\Lambda}\boldsymbol{\Lambda}' + \boldsymbol{\Psi})$$
$$= -\frac{np}{2} \log 2\pi - \frac{n}{2} \log |\boldsymbol{\Lambda}\boldsymbol{\Lambda}' + \boldsymbol{\Psi}| - \frac{n}{2} \operatorname{tr}\left\{ \mathbf{S}(\boldsymbol{\Lambda}\boldsymbol{\Lambda}' + \boldsymbol{\Psi})^{-1} \right\}, \tag{3.3}$$

where

$$\mathbf{S} = \frac{1}{n} \sum_{i=1}^{n} (\mathbf{x}_i - \boldsymbol{\mu})(\mathbf{x}_i - \boldsymbol{\mu})'. \tag{3.4}$$

The maximum likelihood estimate for $\boldsymbol{\mu}$ is easily obtained by differentiating (3.3) with respect to $\boldsymbol{\mu}$ and setting the resulting score function equal to zero to get $\hat{\boldsymbol{\mu}} = \bar{\mathbf{x}}$. An EM algorithm can be used to obtain maximum likelihood estimates for $\boldsymbol{\Lambda}$ and $\boldsymbol{\Psi}$.

3.1.2   An EM Algorithm for the Factor Analysis Model

E-step

The complete-data comprise the observed $\mathbf{x}_1, \ldots, \mathbf{x}_n$ together with the latent factors $\mathbf{u}_1, \ldots, \mathbf{u}_n$, where $\mathbf{u}_i = (u_{i1}, \ldots, u_{iq})'$. Now, noting that $\mathbf{X}_i \mid \mathbf{u}_i \sim N(\boldsymbol{\mu} + \boldsymbol{\Lambda}\mathbf{u}_i, \boldsymbol{\Psi})$, we have

$$\log f(\mathbf{x}_i \mid \mathbf{u}_i) = -\frac{p}{2} \log 2\pi - \frac{1}{2} \log |\boldsymbol{\Psi}| - \frac{1}{2}(\mathbf{x}_i - \boldsymbol{\mu} - \boldsymbol{\Lambda}\mathbf{u}_i)'\boldsymbol{\Psi}^{-1}(\mathbf{x}_i - \boldsymbol{\mu} - \boldsymbol{\Lambda}\mathbf{u}_i)$$
$$= -\frac{p}{2} \log 2\pi - \frac{1}{2} \log |\boldsymbol{\Psi}| - \frac{1}{2}(\mathbf{x}_i - \boldsymbol{\mu})'\boldsymbol{\Psi}^{-1}(\mathbf{x}_i - \boldsymbol{\mu})$$
$$+ (\mathbf{x}_i - \boldsymbol{\mu})'\boldsymbol{\Psi}^{-1}\boldsymbol{\Lambda}\mathbf{u}_i - \frac{1}{2}\mathbf{u}_i'\boldsymbol{\Lambda}'\boldsymbol{\Psi}^{-1}\boldsymbol{\Lambda}\mathbf{u}_i$$
$$= -\frac{p}{2} \log 2\pi - \frac{1}{2} \log |\boldsymbol{\Psi}| - \frac{1}{2} \operatorname{tr}\left\{ \boldsymbol{\Psi}^{-1}(\mathbf{x}_i - \boldsymbol{\mu})(\mathbf{x}_i - \boldsymbol{\mu})' \right\}$$
$$+ (\mathbf{x}_i - \boldsymbol{\mu})'\boldsymbol{\Psi}^{-1}\boldsymbol{\Lambda}\mathbf{u}_i - \frac{1}{2} \operatorname{tr}\left\{ \boldsymbol{\Lambda}'\boldsymbol{\Psi}^{-1}\boldsymbol{\Lambda}\mathbf{u}_i\mathbf{u}_i' \right\}.$$

It follows that the complete-data log-likelihood can be written

$$l_c(\boldsymbol{\mu}, \boldsymbol{\Lambda}, \boldsymbol{\Psi}) = \sum_{i=1}^{n} \log \left[f(\mathbf{x}_i \mid \mathbf{u}_i) f(\mathbf{u}_i) \right]$$
$$= C - \frac{n}{2} \log |\boldsymbol{\Psi}| - \frac{1}{2} \operatorname{tr}\left\{ \boldsymbol{\Psi}^{-1} \sum_{i=1}^{n} (\mathbf{x}_i - \boldsymbol{\mu})(\mathbf{x}_i - \boldsymbol{\mu})' \right\}$$
$$+ \sum_{i=1}^{n} (\mathbf{x}_i - \boldsymbol{\mu})'\boldsymbol{\Psi}^{-1}\boldsymbol{\Lambda}\mathbf{u}_i - \frac{1}{2} \operatorname{tr}\left\{ \boldsymbol{\Lambda}'\boldsymbol{\Psi}^{-1}\boldsymbol{\Lambda} \sum_{i=1}^{n} \mathbf{u}_i\mathbf{u}_i' \right\},$$

where C is constant with respect to $\boldsymbol{\mu}$, $\boldsymbol{\Lambda}$, and $\boldsymbol{\Psi}$.

Consider the joint distribution

$$\begin{bmatrix} \mathbf{X}_i \\ \mathbf{U}_i \end{bmatrix} \sim \mathrm{N}\left(\begin{bmatrix} \boldsymbol{\mu} \\ \mathbf{0} \end{bmatrix}, \begin{bmatrix} \boldsymbol{\Lambda}\boldsymbol{\Lambda}' + \boldsymbol{\Psi} & \boldsymbol{\Lambda} \\ \boldsymbol{\Lambda}' & \mathbf{I}_q \end{bmatrix} \right).$$

It follows that

$$\mathbb{E}[\mathbf{U}_i \mid \mathbf{x}_i] = \boldsymbol{\beta}(\mathbf{x}_i - \boldsymbol{\mu}), \tag{3.5}$$

where $\boldsymbol{\beta} = \boldsymbol{\Lambda}'(\boldsymbol{\Lambda}\boldsymbol{\Lambda}' + \boldsymbol{\Psi})^{-1}$, and

$$\begin{aligned} \mathbb{E}[\mathbf{U}_i\mathbf{U}_i' \mid \mathbf{x}_i] &= \mathrm{Var}[\mathbf{U}_i \mid \mathbf{x}_i] + \mathbb{E}[\mathbf{U}_i \mid \mathbf{x}_i]\mathbb{E}[\mathbf{U}_i \mid \mathbf{x}_i]' \\ &= \mathbf{I}_q - \boldsymbol{\beta}\boldsymbol{\Lambda} + \boldsymbol{\beta}(\mathbf{x}_i - \boldsymbol{\mu})(\mathbf{x}_i - \boldsymbol{\mu})'\boldsymbol{\beta}'. \end{aligned} \tag{3.6}$$

Therefore, noting that $\hat{\boldsymbol{\mu}} = \bar{\mathbf{x}}$ and that we are conditioning on the current parameter estimates, the expected value of the complete-data log-likelihood can be written

$$Q(\boldsymbol{\Lambda}, \boldsymbol{\Psi}) = C - \frac{n}{2}\log|\boldsymbol{\Psi}| - \frac{1}{2}\mathrm{tr}\left\{ \boldsymbol{\Psi}^{-1}\sum_{i=1}^{n}(\mathbf{x}_i - \bar{\mathbf{x}})(\mathbf{x}_i - \bar{\mathbf{x}})' \right\}$$

$$+ \sum_{i=1}^{n}(\mathbf{x}_i - \bar{\mathbf{x}})'\boldsymbol{\Psi}^{-1}\boldsymbol{\Lambda}\,\mathbb{E}\left[\mathbf{U}_i \mid \mathbf{x}_i\right] - \frac{1}{2}\mathrm{tr}\left\{ \boldsymbol{\Lambda}'\boldsymbol{\Psi}^{-1}\boldsymbol{\Lambda}\sum_{i=1}^{n}\mathbb{E}\{\mathbf{U}_i\mathbf{U}_i' \mid \mathbf{x}_i,\} \right\}$$

$$= C + \frac{n}{2}\log|\boldsymbol{\Psi}^{-1}| - \frac{n}{2}\mathrm{tr}\{\boldsymbol{\Psi}^{-1}\mathbf{S}_{\bar{\mathbf{x}}}\} + n\,\mathrm{tr}\{\boldsymbol{\Psi}^{-1}\boldsymbol{\Lambda}\hat{\boldsymbol{\beta}}\mathbf{S}_{\bar{\mathbf{x}}}\} - \frac{n}{2}\mathrm{tr}\{\boldsymbol{\Lambda}'\boldsymbol{\Psi}^{-1}\boldsymbol{\Lambda}\boldsymbol{\Theta}\},$$

where $\boldsymbol{\Theta} = \mathbf{I}_q - \hat{\boldsymbol{\beta}}\hat{\boldsymbol{\Lambda}} + \hat{\boldsymbol{\beta}}\mathbf{S}_{\bar{\mathbf{x}}}\hat{\boldsymbol{\beta}}'$ is a symmetric $q \times q$ matrix, $\hat{\boldsymbol{\beta}} = \hat{\boldsymbol{\Lambda}}'(\hat{\boldsymbol{\Lambda}}\hat{\boldsymbol{\Lambda}}' + \hat{\boldsymbol{\Psi}})^{-1}$, and

$$\mathbf{S}_{\bar{\mathbf{x}}} = \frac{1}{n}\sum_{i=1}^{n}(\mathbf{x}_i - \bar{\mathbf{x}})(\mathbf{x}_i - \bar{\mathbf{x}})' \tag{3.7}$$

can be thought of as the sample, or observed, covariance matrix.

M-step

Differentiating Q with respect to $\boldsymbol{\Lambda}$ and $\boldsymbol{\Psi}$, respectively, gives the score functions

$$\begin{aligned} S_1(\boldsymbol{\Lambda}, \boldsymbol{\Psi}) = \frac{\partial Q}{\partial \boldsymbol{\Lambda}} &= n\frac{\partial}{\partial \boldsymbol{\Lambda}}\mathrm{tr}\{\boldsymbol{\Psi}^{-1}\boldsymbol{\Lambda}\hat{\boldsymbol{\beta}}\mathbf{S}_{\bar{\mathbf{x}}}\} - \frac{n}{2}\frac{\partial}{\partial \boldsymbol{\Lambda}}\mathrm{tr}\{\boldsymbol{\Lambda}'\boldsymbol{\Psi}^{-1}\boldsymbol{\Lambda}\boldsymbol{\Theta}\} \\ &= n(\boldsymbol{\Psi}^{-1})'(\hat{\boldsymbol{\beta}}\mathbf{S}_{\bar{\mathbf{x}}})' - \frac{n}{2}\frac{\partial}{\partial \boldsymbol{\Lambda}}\mathrm{tr}\{\boldsymbol{\Lambda}\boldsymbol{\Theta}\boldsymbol{\Lambda}'\boldsymbol{\Psi}^{-1}\} \\ &= n\boldsymbol{\Psi}^{-1}\mathbf{S}_{\bar{\mathbf{x}}}'\hat{\boldsymbol{\beta}}' - \frac{n}{2}\left[(\boldsymbol{\Psi}^{-1})'\boldsymbol{\Lambda}\boldsymbol{\Theta}' + \boldsymbol{\Psi}^{-1}\boldsymbol{\Lambda}\boldsymbol{\Theta}\right] \\ &= n\boldsymbol{\Psi}^{-1}\mathbf{S}_{\bar{\mathbf{x}}}\hat{\boldsymbol{\beta}}' - n\boldsymbol{\Psi}^{-1}\boldsymbol{\Lambda}\boldsymbol{\Theta}, \end{aligned}$$

and

$$\begin{aligned} S_2(\boldsymbol{\Lambda}, \boldsymbol{\Psi}) = \frac{\partial Q}{\partial \boldsymbol{\Psi}^{-1}} &= \frac{n}{2}\boldsymbol{\Psi} - \frac{n}{2}\mathbf{S}_{\bar{\mathbf{x}}}' + n(\boldsymbol{\Lambda}\hat{\boldsymbol{\beta}}\mathbf{S}_{\bar{\mathbf{x}}})' - \frac{n}{2}(\boldsymbol{\Lambda}')'(\boldsymbol{\Lambda}\boldsymbol{\Theta})' \\ &= \frac{n}{2}\boldsymbol{\Psi} - \frac{n}{2}\mathbf{S}_{\bar{\mathbf{x}}}' + n\boldsymbol{\Lambda}\hat{\boldsymbol{\beta}}\mathbf{S}_{\bar{\mathbf{x}}} - \frac{n}{2}\boldsymbol{\Lambda}\boldsymbol{\Theta}'\boldsymbol{\Lambda}'. \end{aligned}$$

Solving the equations $S_1(\hat{\Lambda}^{\text{new}}, \hat{\Psi}^{\text{new}}) = \mathbf{0}$ and $\text{diag}\{S_2(\hat{\Lambda}^{\text{new}}, \hat{\Psi}^{\text{new}})\} = \mathbf{0}$ gives

$$\hat{\Lambda}^{\text{new}} = \mathbf{S}_{\bar{\mathbf{x}}} \hat{\beta}' \mathbf{\Theta}^{-1},$$

$$\hat{\Psi}^{\text{new}} = \text{diag}\{\mathbf{S}_{\bar{\mathbf{x}}}' - 2\hat{\Lambda}^{\text{new}} \hat{\beta} \mathbf{S}_{\bar{\mathbf{x}}} + \hat{\Lambda}^{\text{new}} \mathbf{\Theta}' (\mathbf{\Theta}^{-1})' \hat{\beta} \mathbf{S}_{\bar{\mathbf{x}}}\}$$

$$= \text{diag}\{\mathbf{S}_{\bar{\mathbf{x}}} - \hat{\Lambda}^{\text{new}} \hat{\beta} \mathbf{S}_{\bar{\mathbf{x}}}\}.$$

The matrix results used to compute these score functions, and used elsewhere in this monograph, are listed in Appendices A.1 and A.2.

The Algorithm

The EM algorithm for the factor analysis model alternates between the E- and M-steps until convergence.

EM Algorithm for Factor Analysis

compute $\mathbf{S}_{\bar{\mathbf{x}}}$
initialize $\hat{\Lambda}$ and $\hat{\Psi}$
while convergence criterion not met
  compute $\hat{\beta} = \hat{\Lambda}'(\hat{\Lambda}\hat{\Lambda}' + \hat{\Psi})^{-1}$
  compute $\mathbf{\Theta} = \mathbf{I}_q - \hat{\beta}\hat{\Lambda} + \hat{\beta}\mathbf{S}_{\bar{\mathbf{x}}}\hat{\beta}'$
  update $\hat{\Lambda}^{\text{new}} = \mathbf{S}_{\bar{\mathbf{x}}}\hat{\beta}'\mathbf{\Theta}^{-1}$
  update $\hat{\Psi}^{\text{new}} = \text{diag}\{\mathbf{S}_{\bar{\mathbf{x}}} - \hat{\Lambda}^{\text{new}}\hat{\beta}\mathbf{S}_{\bar{\mathbf{x}}}\}$
  check convergence criterion
  $\hat{\Lambda} \leftarrow \hat{\Lambda}^{\text{new}}, \hat{\Psi} \leftarrow \hat{\Psi}^{\text{new}}$
end while

The parameters $\hat{\Lambda}$ and $\hat{\Psi}$ can be initialized in several ways, including via eigen-decomposition of $\mathbf{S}_{\bar{\mathbf{x}}}$ (cf. Section 3.3.2). There are various options for convergence criteria for the EM algorithm for the factor analysis model, e.g., a criterion based on Aitken's acceleration could be used (cf. Section 2.2.5).

3.1.3 Woodbury Identity

On each iteration of the EM algorithm for factor analysis, the $p \times p$ matrix $(\hat{\Lambda}\hat{\Lambda}' + \hat{\Psi})^{-1}$ needs to be computed. Computing this matrix inverse can be computationally expensive, especially for larger values of p. The Woodbury identity (Woodbury, 1950) can be used in such situations to avoid inversion of non-diagonal $p \times p$ matrices. For an $m \times m$ matrix $\mathbf{A}$, an $m \times k$ matrix $\mathbf{U}$, a $k \times k$ matrix $\mathbf{C}$, and a $k \times m$ matrix $\mathbf{V}$, the Woodbury identity is

$$(\mathbf{A} + \mathbf{UCV})^{-1} = \mathbf{A}^{-1} - \mathbf{A}^{-1}\mathbf{U}(\mathbf{C}^{-1} + \mathbf{VA}^{-1}\mathbf{U})^{-1}\mathbf{VA}^{-1}.$$

Setting $\mathbf{U} = \Lambda$, $\mathbf{V} = \Lambda'$, $\mathbf{A} = \Psi$, and $\mathbf{C} = \mathbf{I}_q$ gives

$$(\Psi + \Lambda\Lambda')^{-1} = \Psi^{-1} - \Psi^{-1}\Lambda(\mathbf{I}_q + \Lambda'\Psi^{-1}\Lambda)^{-1}\Lambda'\Psi^{-1}, \quad (3.8)$$

which can be used to speed up the EM algorithm for the factor analysis model. The left-hand side of (3.8) requires inversion of a $p \times p$ matrix but the right-hand side leaves only a $q \times q$ matrix and some diagonal matrices to be inverted. A related identity for the determinant of the covariance matrix,

$$|\mathbf{\Lambda}\mathbf{\Lambda}' + \mathbf{\Psi}| = \frac{|\mathbf{\Psi}|}{|\mathbf{I}_q - \mathbf{\Lambda}'(\mathbf{\Lambda}\mathbf{\Lambda}' + \mathbf{\Psi})^{-1}\mathbf{\Lambda}|}, \tag{3.9}$$

is also helpful in computation of the component densities. Identities (3.8) and (3.9) give an especially significant computational advantage when p is large and $q \ll p$.

3.1.4 Comments

The choice of the number of factors is an important consideration in factor analysis. One approach is to choose the number of factors that captures a certain proportion of the variation in the data. Lopes and West (2004) carry out simulation studies showing that the BIC can be effective for selection of the number of factors. Another well-known approach for selecting the number of factors is parallel analysis (Horn, 1965; Humphreys and Ilgen, 1969; Humphreys and Montanelli, 1975; Montanelli and Humphreys, 1976). Different approaches for selecting the number of factors are discussed, *inter alia*, by Fabrigar et al. (1999).

Note that the matrix of factor loadings $\mathbf{\Lambda}$ is not uniquely defined, i.e., if $\mathbf{\Lambda}$ is replaced by $\mathbf{\Lambda}^* = \mathbf{\Lambda}\mathbf{D}$, where $\mathbf{D}$ is a $q \times q$ orthonormal matrix, then

$$\mathbf{\Lambda}\mathbf{\Lambda}' + \mathbf{\Psi} = (\mathbf{\Lambda}^*)(\mathbf{\Lambda}^*)' + \mathbf{\Psi}.$$

Therefore, the factor loadings can be rotated without altering the covariance, i.e., without altering $\mathbf{\Lambda}\mathbf{\Lambda}' + \mathbf{\Psi}$. This is an important consideration when interpreting factor loadings. An elegant geometric interpretation is given by Lawley and Maxwell (1962), who explain that, in a sense, the q-dimensional space in which the factors lie can be determined but the directions of the factors in that space cannot.

3.2 Mixture of Factor Analyzers

Analogous to the factor analysis model, the mixture of factor analyzers model assumes that

$$\mathbf{X}_i = \boldsymbol{\mu}_g + \mathbf{\Lambda}_g \mathbf{U}_{ig} + \boldsymbol{\varepsilon}_{ig} \tag{3.10}$$

with probability π_g, for $i = 1, \ldots, n$ and $g = 1, \ldots, G$, where $\mathbf{\Lambda}_g$ is a $p \times q$ matrix of factor loadings, the $\mathbf{U}_{ig}$ are independently $N(\mathbf{0}, \mathbf{I}_q)$ and are independent of the $\boldsymbol{\varepsilon}_{ig}$, which are independently $N(\mathbf{0}, \mathbf{\Psi}_g)$, where $\mathbf{\Psi}_g$ is a $p \times p$

diagonal matrix with positive diagonal elements. It follows that the density of $\mathbf{X}_i$ from the mixture of factor analyzers model is

$$f(\mathbf{x}_i \mid \boldsymbol{\vartheta}) = \sum_{g=1}^{G} \pi_g \phi(\mathbf{x}_i \mid \boldsymbol{\mu}_g, \boldsymbol{\Lambda}_g \boldsymbol{\Lambda}_g' + \boldsymbol{\Psi}_g), \qquad (3.11)$$

where $\boldsymbol{\vartheta}$ denotes the model parameters.

Ghahramani and Hinton (1997) were the first to introduce a mixture of factor analyzers model. In their model, they constrain $\boldsymbol{\Psi}_g = \boldsymbol{\Psi}$ to facilitate an interpretation of $\boldsymbol{\Psi}$ as sensor noise; however, they note that it is possible to relax this constraint. Tipping and Bishop (1997, 1999) introduce the closely related mixture of probabilistic principal component analyzers (MPPCA) model, where each $\boldsymbol{\Psi}_g$ matrix is isotropic, i.e., $\boldsymbol{\Psi}_g = \psi_g \mathbf{I}_p$, so that $\boldsymbol{\Sigma}_g = \boldsymbol{\Lambda}_g \boldsymbol{\Lambda}_g' + \psi_g \mathbf{I}_p$. McLachlan and Peel (2000b) use the unconstrained mixture of factor analyzers model, i.e., with $\boldsymbol{\Sigma}_g = \boldsymbol{\Lambda}_g \boldsymbol{\Lambda}_g' + \boldsymbol{\Psi}_g$.

3.3   Parsimonious Gaussian Mixture Models

3.3.1   A Family of Eight Models

One can view the mixture of factor analyzers models and the MPPCA model, collectively, as a family of three models, where two members arise from imposing constraints on the most general model, i.e., the model with $\boldsymbol{\Sigma}_g = \boldsymbol{\Lambda}_g \boldsymbol{\Lambda}_g' + \boldsymbol{\Psi}_g$ (cf. Section 3.2). This family can easily be extended to a four-member family by adding the model with component covariance $\boldsymbol{\Sigma}_g = \boldsymbol{\Lambda}_g \boldsymbol{\Lambda}_g' + \psi \mathbf{I}_p$. Members of this family of four models have between $G[pq - q(q-1)/2] + 1$ and $G[pq - q(q-1)/2] + Gp$ free parameters in the component covariance matrices. A greater level of parsimony can be introduced by constraining the component factor loading matrices to be equal, i.e., $\boldsymbol{\Lambda}_g = \boldsymbol{\Lambda}$. McNicholas and Murphy (2005, 2008) develop a family of eight parsimonious Gaussian mixture models (PGMMs) for clustering by imposing, or not, each of the constraints $\boldsymbol{\Lambda}_g = \boldsymbol{\Lambda}$, $\boldsymbol{\Psi}_g = \boldsymbol{\Psi}$, and $\boldsymbol{\Psi}_g = \psi_g \mathbf{I}_p$. Members of the PGMM family have between $pq - q(q-1)/2 + 1$ and $G[pq - q(q-1)/2] + Gp$ free parameters in the component covariance matrices (cf. Table 3.1). McNicholas (2010) used the PGMM family for model-based classification, and Andrews and McNicholas (2011b) applied it for model-based discriminant analysis.

3.3.2   Parameter Estimation

Parameter estimation for members of the PGMM family can be carried out using alternating expectation-conditional maximization (AECM) algorithms (Meng and van Dyk, 1997). The expectation-conditional maximization (ECM)

Table 3.1

The nomenclature, covariance structure, and number of free covariance parameters for each member of the PGMM family, where "C" denotes "constrained", i.e., the constraint is imposed, and "U" denotes "unconstrained", i.e., the constraint is not imposed.

$\Lambda_g = \Lambda$	$\Psi_g = \Psi$	$\Psi_g = \psi_g \mathbf{I}_p$	Σ_g	Free Cov. Paras.
C	C	C	$\Lambda\Lambda' + \psi\mathbf{I}_p$	$pq - q(q-1)/2 + 1$
C	C	U	$\Lambda\Lambda' + \Psi$	$pq - q(q-1)/2 + p$
C	U	C	$\Lambda\Lambda' + \psi_g\mathbf{I}_p$	$pq - q(q-1)/2 + G$
C	U	U	$\Lambda\Lambda' + \Psi_g$	$pq - q(q-1)/2 + Gp$
U	C	C	$\Lambda_g\Lambda_g' + \psi\mathbf{I}_p$	$G[pq - q(q-1)/2] + 1$
U	C	U	$\Lambda_g\Lambda_g' + \Psi$	$G[pq - q(q-1)/2] + p$
U	U	C	$\Lambda_g\Lambda_g' + \psi_g\mathbf{I}_p$	$G[pq - q(q-1)/2] + G$
U	U	U	$\Lambda_g\Lambda_g' + \Psi_g$	$G[pq - q(q-1)/2] + Gp$

algorithm (Meng and Rubin, 1993) is a variant of the EM algorithm that replaces the M-step by a series of conditional maximization steps. The AECM algorithm allows a different specification of complete-data for each conditional maximization step. This makes it a convenient approach for the PGMM models, where there are two sources of missing data: the unknown component membership labels z_{ig} and the latent factors $\mathbf{u}_{ig}$, for $i = 1, \ldots, n$ and $g = 1, \ldots, G$. Details of fitting the AECM algorithm for the more general mixture of factor analyzers model are given by McLachlan and Peel (2000b), and parameter estimation for other members of the PGMM family is discussed by McNicholas and Murphy (2008) and McNicholas et al. (2010).

AECM Algorithm: First Stage

As usual, denote by $\mathbf{z}_1, \ldots, \mathbf{z}_n$ the unobserved group membership labels, where $z_{ig} = 1$ if observation i belongs to component g and $z_{ig} = 0$ otherwise. At the first stage of the AECM algorithm, the complete-data are taken to be the observed $\mathbf{x}_1, \ldots, \mathbf{x}_n$ together with the unobserved $\mathbf{z}_1, \ldots, \mathbf{z}_n$, and the parameters π_g and $\boldsymbol{\mu}_g$ are estimated, for $g = 1, \ldots, G$. The complete-data log-likelihood is

$$l_1 = \sum_{i=1}^{n}\sum_{g=1}^{G} z_{ig} \log\left[\pi_g \phi(\mathbf{x}_i \mid \boldsymbol{\mu}_g, \Lambda_g\Lambda_g' + \Psi_g)\right], \qquad (3.12)$$

and the (conditional) expected values of the component membership labels are given by

$$\hat{z}_{ig} = \frac{\hat{\pi}_g \phi(\mathbf{x}_i \mid \hat{\boldsymbol{\mu}}_g, \hat{\Lambda}_g\hat{\Lambda}_g' + \hat{\Psi}_g)}{\sum_{h=1}^{G} \hat{\pi}_h \phi(\mathbf{x}_i \mid \hat{\boldsymbol{\mu}}_h, \hat{\Lambda}_h\hat{\Lambda}_h' + \hat{\Psi}_h)}, \qquad (3.13)$$

for $i = 1, \ldots, n$ and $g = 1, \ldots, G$.

Using the expected values given by (3.13) in (3.12), the expected value of the complete-data log-likelihood at the first stage is

$$Q_1 = \sum_{i=1}^{n} \sum_{g=1}^{G} \hat{z}_{ig} \left[\log \pi_g + \log \phi(\mathbf{x}_i \mid \boldsymbol{\mu}_g, \boldsymbol{\Lambda}_g \boldsymbol{\Lambda}'_g + \boldsymbol{\Psi}_g) \right]$$

$$= \sum_{g=1}^{G} n_g \log \pi_g - \frac{np}{2} \log 2\pi - \sum_{g=1}^{G} \frac{n_g}{2} \log |\boldsymbol{\Lambda}_g \boldsymbol{\Lambda}'_g + \boldsymbol{\Psi}_g|$$

$$- \sum_{g=1}^{G} \frac{n_g}{2} \operatorname{tr} \left\{ \mathbf{S}_g (\boldsymbol{\Lambda}_g \boldsymbol{\Lambda}'_g + \boldsymbol{\Psi}_g)^{-1} \right\},$$

where $n_g = \sum_{i=1}^{n} \hat{z}_{ig}$ and

$$\mathbf{S}_g = \frac{1}{n_g} \sum_{i=1}^{n} \hat{z}_{ig} (\mathbf{x}_i - \boldsymbol{\mu}_g)(\mathbf{x}_i - \boldsymbol{\mu}_g)'. \tag{3.14}$$

Maximising Q_1 with respect to π_g and $\boldsymbol{\mu}_g$ yields

$$\hat{\pi}_g = \frac{n_g}{n} \qquad \text{and} \qquad \hat{\boldsymbol{\mu}}_g = \frac{\sum_{i=1}^{n} \hat{z}_{ig} \mathbf{x}_i}{\sum_{i=1}^{n} \hat{z}_{ig}}, \tag{3.15}$$

respectively.

AECM Algorithm: Second Stage

At the second stage of the AECM algorithm, the complete-data are taken to be the observed $\mathbf{x}_1, \ldots, \mathbf{x}_n$ together with the unobserved component membership labels $\mathbf{z}_1, \ldots, \mathbf{z}_n$ and the latent factors $\mathbf{u}_{ig}$, for $i = 1, \ldots, n$ and $g = 1, \ldots, G$, and the parameters $\boldsymbol{\Lambda}_g$ and $\boldsymbol{\Psi}_g$ are estimated, for $g = 1, \ldots, G$. Proceeding in an analogous fashion to the EM algorithm for the factor analysis model (Section 3.1.2), the complete-data log-likelihood is given by

$$l_2 = \sum_{i=1}^{n} \sum_{g=1}^{G} \hat{z}_{ig} \left[\log \pi_g + \log f(\mathbf{x}_i | \mathbf{u}_i) + \log f(\mathbf{u}_i) \right]$$

$$= C + \sum_{g=1}^{G} \left[-\frac{n_g}{2} \log |\boldsymbol{\Psi}_g| - \frac{n_g}{2} \operatorname{tr} \left\{ \boldsymbol{\Psi}_g^{-1} \mathbf{S}_g \right\} + \sum_{i=1}^{n} z_{ig} (\mathbf{x}_i - \boldsymbol{\mu}_g)' \boldsymbol{\Psi}_g^{-1} \boldsymbol{\Lambda}_g \mathbf{u}_i \right.$$

$$\left. - \frac{1}{2} \operatorname{tr} \left\{ \boldsymbol{\Lambda}'_g \boldsymbol{\Psi}_g^{-1} \boldsymbol{\Lambda}_g \sum_{i=1}^{n} z_{ig} \mathbf{u}_i \mathbf{u}'_i \right\} \right],$$

where C is constant with respect to $\boldsymbol{\Lambda}_g$ and $\boldsymbol{\Psi}_g$. Bearing in mind that we are conditioning on the current parameter estimates, and using expected values analogous to those in (3.5) and (3.6), the expected value of the complete-data log-likelihood can be written

$$
Q_2 = C + \sum_{g=1}^{G} \left[-\frac{n_g}{2} \log |\boldsymbol{\Psi}_g| - \frac{n_g}{2} \operatorname{tr}\left\{ \boldsymbol{\Psi}_g^{-1} \mathbf{S}_g \right\} \right.
$$

$$
+ \sum_{i=1}^{n} \hat{z}_{ig} (\mathbf{x}_i - \hat{\boldsymbol{\mu}}_g)' \boldsymbol{\Psi}_g^{-1} \boldsymbol{\Lambda}_g \mathbb{E}[\mathbf{U}_{ig} \mid \mathbf{x}_i, z_{ig} = 1]
$$

$$
\left. - \frac{1}{2} \operatorname{tr}\left\{ \boldsymbol{\Lambda}_g' \boldsymbol{\Psi}_g^{-1} \boldsymbol{\Lambda}_g \sum_{i=1}^{n} \hat{z}_{ig} \mathbb{E}[\mathbf{U}_{ig} \mathbf{U}_{ig}' \mid \mathbf{x}_i, z_{ig} = 1] \right\} \right]
$$

$$
= C + \frac{1}{2} \sum_{g=1}^{G} n_g \left[\log |\boldsymbol{\Psi}_g^{-1}| - \operatorname{tr}\left\{ \boldsymbol{\Psi}_g^{-1} \mathbf{S}_g \right\} + 2 \operatorname{tr}\left\{ \boldsymbol{\Psi}_g^{-1} \boldsymbol{\Lambda}_g \hat{\boldsymbol{\beta}}_g \mathbf{S}_g \right\} \right.
$$

$$
\left. - \operatorname{tr}\left\{ \boldsymbol{\Lambda}_g' \boldsymbol{\Psi}_g^{-1} \boldsymbol{\Lambda}_g \boldsymbol{\Theta}_g \right\} \right],
$$

where $\hat{\boldsymbol{\beta}}_g = \hat{\boldsymbol{\Lambda}}_g' (\hat{\boldsymbol{\Lambda}}_g \hat{\boldsymbol{\Lambda}}_g' + \hat{\boldsymbol{\Psi}}_g)^{-1}$ and $\boldsymbol{\Theta}_g = \mathbf{I}_q - \hat{\boldsymbol{\beta}}_g \hat{\boldsymbol{\Lambda}}_g + \hat{\boldsymbol{\beta}}_g \mathbf{S}_g \hat{\boldsymbol{\beta}}_g'$. Note that $\hat{\boldsymbol{\mu}}_g$ replaces $\boldsymbol{\mu}_g$ in $\mathbf{S}_g$, cf. (3.14).

For all but the UUU model, i.e., the mixture of factor analyzers model, constraints will be imposed on $\boldsymbol{\Lambda}_g$ and/or $\boldsymbol{\Psi}_g$, with knock-on consequences for $\boldsymbol{\beta}_g$, $\boldsymbol{\Theta}_g$, and Q_2. The nature of these constraints will depend on the model under consideration (Table 3.1). Consider the CCU model, so that $\boldsymbol{\Lambda}_g = \boldsymbol{\Lambda}$ and $\boldsymbol{\Psi}_g = \boldsymbol{\Psi}$, and Q_2 can be written

$$
Q_2(\boldsymbol{\Lambda}, \boldsymbol{\Psi}) =
$$
$$
C + \frac{n}{2} \left[\log |\boldsymbol{\Psi}^{-1}| - \operatorname{tr}\left\{ \boldsymbol{\Psi}^{-1} \tilde{\mathbf{S}} \right\} + 2 \operatorname{tr}\left\{ \boldsymbol{\Psi}^{-1} \boldsymbol{\Lambda} \hat{\boldsymbol{\beta}} \tilde{\mathbf{S}} \right\} - \operatorname{tr}\left\{ \boldsymbol{\Lambda}' \boldsymbol{\Psi}^{-1} \boldsymbol{\Lambda} \tilde{\boldsymbol{\Theta}} \right\} \right],
$$

where $\tilde{\mathbf{S}} = \sum_{g=1}^{G} \hat{\pi}_g \mathbf{S}_g$, $\tilde{\boldsymbol{\Theta}} = \mathbf{I}_q - \hat{\boldsymbol{\beta}} \hat{\boldsymbol{\Lambda}} + \hat{\boldsymbol{\beta}} \tilde{\mathbf{S}} \hat{\boldsymbol{\beta}}'$, and $\hat{\boldsymbol{\beta}} = \hat{\boldsymbol{\Lambda}}' (\hat{\boldsymbol{\Lambda}} \hat{\boldsymbol{\Lambda}}' + \hat{\boldsymbol{\Psi}})^{-1}$. Differentiating $Q_2(\boldsymbol{\Lambda}, \boldsymbol{\Psi})$ with respect to $\boldsymbol{\Lambda}$ and $\boldsymbol{\Psi}^{-1}$, respectively, gives the score functions

$$
S_1(\boldsymbol{\Lambda}, \boldsymbol{\Psi}) = \frac{\partial Q_2(\boldsymbol{\Lambda}, \boldsymbol{\Psi})}{\partial \boldsymbol{\Lambda}} = n \left[\boldsymbol{\Psi}^{-1} \tilde{\mathbf{S}} \hat{\boldsymbol{\beta}}' - \boldsymbol{\Psi}^{-1} \boldsymbol{\Lambda} \tilde{\boldsymbol{\Theta}} \right],
$$

$$
S_2(\boldsymbol{\Lambda}, \boldsymbol{\Psi}) = \frac{\partial Q_2(\boldsymbol{\Lambda}, \boldsymbol{\Psi})}{\partial \boldsymbol{\Psi}^{-1}} = \frac{n}{2} \left[\boldsymbol{\Psi} - \tilde{\mathbf{S}}' + 2\boldsymbol{\Lambda} \hat{\boldsymbol{\beta}} \tilde{\mathbf{S}} - \boldsymbol{\Lambda} \tilde{\boldsymbol{\Theta}}' \boldsymbol{\Lambda}' \right].
$$

Solving $S_1(\hat{\boldsymbol{\Lambda}}^{\text{new}}, \hat{\boldsymbol{\Psi}}^{\text{new}}) = \mathbf{0}$ and $\operatorname{diag}\{S_2(\hat{\boldsymbol{\Lambda}}^{\text{new}}, \hat{\boldsymbol{\Psi}}^{\text{new}})\} = \mathbf{0}$ gives

$$
\hat{\boldsymbol{\Lambda}}^{\text{new}} = \tilde{\mathbf{S}} \hat{\boldsymbol{\beta}}' \tilde{\boldsymbol{\Theta}}^{-1},
$$

$$
\hat{\boldsymbol{\Psi}}^{\text{new}} = \operatorname{diag}\left\{ \tilde{\mathbf{S}} - 2\hat{\boldsymbol{\Lambda}}^{\text{new}} \hat{\boldsymbol{\beta}} \tilde{\mathbf{S}} + \hat{\boldsymbol{\Lambda}}^{\text{new}} \tilde{\boldsymbol{\Theta}}' (\tilde{\mathbf{S}} \hat{\boldsymbol{\beta}}' \tilde{\boldsymbol{\Theta}}^{-1})' \right\}
$$

$$
= \operatorname{diag}\left\{ \tilde{\mathbf{S}} - \hat{\boldsymbol{\Lambda}}^{\text{new}} \hat{\boldsymbol{\beta}} \tilde{\mathbf{S}} \right\}.
$$

AECM Algorithm for the CCU Model

An AECM algorithm for the CCU model can now be presented.

AECM Algorithm for PGMM Model CCU

initialize $\hat{z}_{ig}$
initialize $\hat{\pi}_g, \hat{\boldsymbol{\mu}}_g, \tilde{\mathbf{S}}, \hat{\boldsymbol{\Lambda}}, \hat{\boldsymbol{\Psi}}$
while convergence criterion not met
    update $\hat{\pi}_g, \hat{\boldsymbol{\mu}}_g$
    if not iteration 1
        update $\hat{z}_{ig}$
    end if
    compute $\tilde{\mathbf{S}}, \hat{\beta}, \tilde{\boldsymbol{\Theta}}$
    update $\hat{\boldsymbol{\Lambda}}^{\text{new}}, \hat{\boldsymbol{\Psi}}^{\text{new}}$
    update $\hat{z}_{ig}$
    check convergence criterion
    $\hat{\boldsymbol{\Lambda}} \leftarrow \hat{\boldsymbol{\Lambda}}^{\text{new}}, \hat{\boldsymbol{\Psi}} \leftarrow \hat{\boldsymbol{\Psi}}^{\text{new}}$
end while

There are a few different options for initializing the $\hat{z}_{ig}$, and k-means clustering is the default in the `pgmm` package for R. Based on the initial values for the $\hat{z}_{ig}$, $\hat{\pi}_g$ and $\hat{\boldsymbol{\mu}}_g$ can be initialized based on their updates, cf. (3.15). Finding suitable initial values for $\hat{\boldsymbol{\Lambda}}$ and $\hat{\boldsymbol{\Psi}}$ is a little more difficult. McNicholas and Murphy (2008) suggest an approach based on the eigen-decomposition of $\tilde{\mathbf{S}}$. Specifically, $\tilde{\mathbf{S}}$ is computed based on the initial values of the $\hat{z}_{ig}$. Then the eigen-decomposition of $\tilde{\mathbf{S}}$ is computed to give

$$\tilde{\mathbf{S}} = \mathbf{PDP}^{-1},$$

and $\hat{\boldsymbol{\Lambda}}$ is initialized using

$$\hat{\boldsymbol{\Lambda}} = \mathbf{dP},$$

where $\mathbf{d}$ is the element-wise square root of the diagonal of $\mathbf{D}$. The initial value of $\boldsymbol{\Psi}$ is taken to be

$$\hat{\boldsymbol{\Psi}} = \text{diag}\{\tilde{\mathbf{S}} - (\mathbf{dP})(\mathbf{dP})'\}.$$

Analogous approaches can be taken to find initial values for the covariance parameters in the other members of the PGMM family (Table 3.1), and examples can be found in McNicholas and Murphy (2008) and McNicholas et al. (2010). There are several different options for convergence criteria for the AECM algorithm for the PGMM models. A criterion based on Aitken's acceleration, cf. Section 2.2.5, is used in `pgmm`.

3.3.3 Comments

Note the analogy between the AECM algorithms presented for the PGMM models and the EM algorithm for the factor analysis model (Section 3.1.2). This is most apparent when one considers the AECM algorithm for the CCU model when $G = 1$; this algorithm is effectively the same as the EM algorithm for the factor analysis model presented in Section 3.1.2.

3.4 Expanded Parsimonious Gaussian Mixture Models

3.4.1 A Family of Twelve Models

McNicholas and Murphy (2010b) further parameterize the mixture of factor analyzers component covariance structure by writing $\boldsymbol{\Psi}_g = \omega_g \boldsymbol{\Delta}_g$, where $\omega_g \in \mathbb{R}^+$ and $\boldsymbol{\Delta}_g$ is a diagonal matrix with $|\boldsymbol{\Delta}_g| = 1$. The resulting mixture of modified factor analyzers model has component covariance structure

$$\boldsymbol{\Sigma}_g = \boldsymbol{\Lambda}_g \boldsymbol{\Lambda}'_g + \omega_g \boldsymbol{\Delta}_g.$$

In addition to the constraint $\boldsymbol{\Lambda}_g = \boldsymbol{\Lambda}$, all legitimate combinations of the constraints $\omega_g = \omega$, $\boldsymbol{\Delta}_g = \boldsymbol{\Delta}$, and $\boldsymbol{\Delta}_g = \mathbf{I}_p$ are imposed, resulting in a family of 12 parsimonious Gaussian mixture models (Table 3.2). For clarity, this family will be called the expanded PGMM family for the remainder of this section.

Table 3.2
The covariance structure and nomenclature for each member of the expanded PGMM family, along with the name of the equivalent member of the PGMM family (Table 3.1), where applicable.

Expanded PGMM Nomenclature					
$\boldsymbol{\Lambda}_g = \boldsymbol{\Lambda}$	$\boldsymbol{\Delta}_g = \boldsymbol{\Delta}$	$\omega_g = \omega$	$\boldsymbol{\Delta}_g = \mathbf{I}_p$	PGMM Equiv.	$\boldsymbol{\Sigma}_g$
C	C	C	C	CCC	$\boldsymbol{\Lambda}\boldsymbol{\Lambda}' + \omega \mathbf{I}_p$
C	C	U	C	CUC	$\boldsymbol{\Lambda}\boldsymbol{\Lambda}' + \omega_g \mathbf{I}_p$
U	C	C	C	UCC	$\boldsymbol{\Lambda}_g \boldsymbol{\Lambda}'_g + \omega \mathbf{I}_p$
U	C	U	C	UUC	$\boldsymbol{\Lambda}_g \boldsymbol{\Lambda}'_g + \omega_g \mathbf{I}_p$
C	C	C	U	CCU	$\boldsymbol{\Lambda}\boldsymbol{\Lambda}' + \omega \boldsymbol{\Delta}$
C	C	U	U	–	$\boldsymbol{\Lambda}\boldsymbol{\Lambda}' + \omega_g \boldsymbol{\Delta}$
U	C	C	U	UCU	$\boldsymbol{\Lambda}_g \boldsymbol{\Lambda}'_g + \omega \boldsymbol{\Delta}$
U	C	U	U	–	$\boldsymbol{\Lambda}_g \boldsymbol{\Lambda}'_g + \omega_g \boldsymbol{\Delta}$
C	U	C	U	–	$\boldsymbol{\Lambda}\boldsymbol{\Lambda}' + \omega \boldsymbol{\Delta}_g$
C	U	U	U	CUU	$\boldsymbol{\Lambda}\boldsymbol{\Lambda}' + \omega_g \boldsymbol{\Delta}_g$
U	U	C	U	–	$\boldsymbol{\Lambda}_g \boldsymbol{\Lambda}'_g + \omega \boldsymbol{\Delta}_g$
U	U	U	U	UUU	$\boldsymbol{\Lambda}_g \boldsymbol{\Lambda}'_g + \omega_g \boldsymbol{\Delta}_g$

The pgmm package for R implements all 12 PGMM models for model-based clustering and classification. Similar to the PGMM family, a key feature of the expanded PGMM family is that all members have $\mathcal{O}(p)$ covariance parameters, i.e., the number of covariance parameters is linear in the dimensionality of the data. This is very important in the analysis of high-dimensional data.

3.4.2   Parameter Estimation

Parameter estimation for the expanded PGMM family proceeds in a very similar fashion as for the PGMM family (Section 3.3.2). In fact, updates for the eight extended PGMM models that have PGMM analogues (Table 3.2) can be obtained directly from the analogous PGMM updates via

$$\hat{\omega}_g = |\hat{\boldsymbol{\Psi}}_g|^{1/p},$$
$$\hat{\boldsymbol{\Delta}}_g = \hat{\boldsymbol{\Psi}}_g/|\hat{\boldsymbol{\Psi}}_g|^{1/p}.$$

For example, consider the CCCU model. This model is equivalent to the CCU model considered in Section 3.3.2, where

$$\hat{\boldsymbol{\Lambda}}^{\text{new}} = \tilde{\mathbf{S}}\hat{\boldsymbol{\beta}}'\tilde{\boldsymbol{\Theta}}^{-1},$$
$$\hat{\boldsymbol{\Psi}}^{\text{new}} = \text{diag}\{\tilde{\mathbf{S}} - \hat{\boldsymbol{\Lambda}}^{\text{new}}\hat{\boldsymbol{\beta}}\tilde{\mathbf{S}}\}.$$

It then follows that the updates for the CCCU model are given by

$$\hat{\boldsymbol{\Lambda}}^{\text{new}} = \tilde{\mathbf{S}}\hat{\boldsymbol{\beta}}'\tilde{\boldsymbol{\Theta}}^{-1},$$
$$\hat{\omega}^{\text{new}} = |\text{diag}\{\tilde{\mathbf{S}} - \hat{\boldsymbol{\Lambda}}^{\text{new}}\hat{\boldsymbol{\beta}}\tilde{\mathbf{S}}\}|^{1/p},$$
$$\hat{\boldsymbol{\Delta}}^{\text{new}} = \frac{1}{|\text{diag}\{\tilde{\mathbf{S}} - \hat{\boldsymbol{\Lambda}}^{\text{new}}\hat{\boldsymbol{\beta}}\tilde{\mathbf{S}}\}|^{1/p}} \, \text{diag}\{\tilde{\mathbf{S}} - \hat{\boldsymbol{\Lambda}}^{\text{new}}\hat{\boldsymbol{\beta}}\tilde{\mathbf{S}}\},$$

where $\hat{\boldsymbol{\beta}} = \hat{\boldsymbol{\Lambda}}'(\hat{\boldsymbol{\Lambda}}\hat{\boldsymbol{\Lambda}}' + \hat{\omega}\hat{\boldsymbol{\Delta}})^{-1}$, and $\tilde{\boldsymbol{\Theta}}$ and $\tilde{\mathbf{S}}$ are as defined in Section 3.3.2.

The presence of the constraint $|\boldsymbol{\Delta}_g| = 1$, however, makes the second stage of the AECM for the extended PGMM family a little more complicated for the four models that do not have a PGMM analogue. In these cases, McNicholas and Murphy (2010b) use the method of Lagrange multipliers (cf. Appendix A.3). For illustration, consider the UCUU model so that $\boldsymbol{\Delta}_g = \boldsymbol{\Delta}$ and, at the second stage of the AECM algorithm, the expected value of the complete-data log-likelihood is

$$Q_2 = C + \frac{1}{2}\sum_{g=1}^{G} n_g \left[p\log\omega_g^{-1} + \log|\boldsymbol{\Delta}^{-1}| - \omega_g^{-1}\text{tr}\{\boldsymbol{\Delta}^{-1}\mathbf{S}_g\} \right.$$
$$\left. + 2\omega_g^{-1}\text{tr}\{\boldsymbol{\Delta}^{-1}\boldsymbol{\Lambda}_g\hat{\boldsymbol{\beta}}_g\mathbf{S}_g\} - \omega_g^{-1}\text{tr}\{\boldsymbol{\Lambda}_g'\boldsymbol{\Delta}^{-1}\boldsymbol{\Lambda}_g\boldsymbol{\Theta}_g\} \right],$$

where $\hat{\boldsymbol{\beta}}_g = \hat{\boldsymbol{\Lambda}}_g'(\hat{\boldsymbol{\Lambda}}_g\hat{\boldsymbol{\Lambda}}_g' + \hat{\omega}_g\boldsymbol{\Delta})^{-1}$ and $\boldsymbol{\Theta}_g = \mathbf{I}_q - \hat{\boldsymbol{\beta}}_g\hat{\boldsymbol{\Lambda}}_g + \hat{\boldsymbol{\beta}}_g\mathbf{S}_g\hat{\boldsymbol{\beta}}_g'$. Now form the Lagrangian

$$L(\boldsymbol{\Lambda}_g, \omega_g, \boldsymbol{\Delta}, \lambda) = Q_2(\boldsymbol{\Lambda}_g, \omega_g, \boldsymbol{\Delta}) - \iota(|\boldsymbol{\Delta}| - 1),$$

where ι is used to denote the Lagrange multiplier. Differentiating L with respect to $\mathbf{\Lambda}_g$, ω_g^{-1}, $\mathbf{\Delta}^{-1}$, and ι, respectively, gives the score functions

$$S_1(\mathbf{\Lambda}_g, \omega_g, \mathbf{\Delta}, \iota) = \frac{\partial L}{\partial \mathbf{\Lambda}} = \sum_{g=1}^{G} \frac{n_g}{\omega_g} \left[\mathbf{\Delta}^{-1} \mathbf{S}_g \hat{\boldsymbol{\beta}}_g' - \mathbf{\Delta}^{-1} \mathbf{\Lambda}_g \tilde{\mathbf{\Theta}}_g \right],$$

$$S_2(\mathbf{\Lambda}_g, \omega_g, \mathbf{\Delta}, \iota) = \frac{\partial L}{\partial \omega_g^{-1}}$$
$$= \frac{n_g}{2} \left[p\omega_g - \mathrm{tr}\{\mathbf{\Delta}^{-1} \mathbf{S}_g\} + 2\,\mathrm{tr}\{\mathbf{\Delta}^{-1} \mathbf{\Lambda}_g \hat{\boldsymbol{\beta}}_g \mathbf{S}_g\} - \mathrm{tr}\{\mathbf{\Lambda}_g' \mathbf{\Delta}^{-1} \mathbf{\Lambda}_g \mathbf{\Theta}_g\} \right],$$

$$S_3(\mathbf{\Lambda}_g, \omega_g, \mathbf{\Delta}, \iota) = \frac{\partial L}{\partial \mathbf{\Delta}^{-1}}$$
$$= \frac{1}{2} \sum_{g=1}^{G} n_g \left[\mathbf{\Delta} - \omega_g^{-1} \mathbf{S}_g' + 2\omega_g^{-1} \mathbf{\Lambda}_g \hat{\boldsymbol{\beta}}_g \mathbf{S}_g - \omega_g^{-1} \mathbf{\Lambda}_g \mathbf{\Theta}_g' \mathbf{\Lambda}_g' \right] + \lambda |\mathbf{\Delta}| \mathbf{\Delta},$$

$$S_4(\mathbf{\Lambda}_g, \omega_g, \mathbf{\Delta}, \iota) = \frac{\partial L}{\partial \iota} = |\mathbf{\Delta}| - 1.$$

Note that setting S_4 equal to zero just leads to the constraint $|\mathbf{\Delta}| = 1$, and S_4 is included only for completeness. Solving $S_1(\hat{\mathbf{\Lambda}}_g^{\mathrm{new}}, \hat{\omega}_g, \hat{\mathbf{\Delta}}, \iota) = \mathbf{0}$ gives

$$\hat{\mathbf{\Lambda}}_g^{\mathrm{new}} = \mathbf{S}_g \hat{\boldsymbol{\beta}}_g' \mathbf{\Theta}_g^{-1},$$

and solving $S_2(\hat{\mathbf{\Lambda}}_g^{\mathrm{new}}, \hat{\omega}_g^{\mathrm{new}}, \hat{\mathbf{\Delta}}, \iota) = 0$ gives

$$\hat{\omega}_g^{\mathrm{new}} = \frac{1}{p} \mathrm{tr}\{\hat{\mathbf{\Delta}}^{-1} \mathbf{S}_g - \hat{\mathbf{\Delta}}^{-1} \hat{\mathbf{\Lambda}}_g^{\mathrm{new}} \hat{\boldsymbol{\beta}}_g \mathbf{S}_g\}.$$

Now, solving $\mathrm{diag}\{ S_3(\hat{\mathbf{\Lambda}}_g^{\mathrm{new}}, \hat{\omega}_g^{\mathrm{new}}, \hat{\mathbf{\Delta}}^{\mathrm{new}}, \iota) \} = \mathbf{0}$ gives

$$\hat{\mathbf{\Delta}}^{\mathrm{new}} = \frac{1}{n + 2\iota |\hat{\mathbf{\Delta}}^{\mathrm{new}}|} \sum_{g=1}^{G} \frac{n_g}{\hat{\omega}_g^{\mathrm{new}}} \mathrm{diag}\{\mathbf{S}_g - \hat{\mathbf{\Lambda}}_g^{\mathrm{new}} \hat{\boldsymbol{\beta}}_g \mathbf{S}_g\}$$
$$= \frac{1}{n + 2\iota} \mathrm{diag} \left\{ \sum_{g=1}^{G} \frac{n_g}{\hat{\omega}_g^{\mathrm{new}}} \left[\mathbf{S}_g - \hat{\mathbf{\Lambda}}_g^{\mathrm{new}} \hat{\boldsymbol{\beta}}_g \mathbf{S}_g \right] \right\}.$$

Because $\hat{\mathbf{\Delta}}^{\mathrm{new}}$ is a diagonal matrix with $|\hat{\mathbf{\Delta}}^{\mathrm{new}}| = 1$,

$$n + 2\iota = \left(\prod_{i=1}^{p} \xi_j \right)^{\frac{1}{p}},$$

where ξ_j is the jth element along the diagonal of the matrix

$$\sum_{g=1}^{G} \frac{n_g}{\hat{\omega}_g^{\mathrm{new}}} \left[\mathbf{S}_g - \hat{\mathbf{\Lambda}}_g^{\mathrm{new}} \hat{\boldsymbol{\beta}}_g \mathbf{S}_g \right];$$

accordingly,

$$\iota = \frac{1}{2}\left[\left(\prod_{j=1}^{p} \xi_j\right)^{\frac{1}{p}} - n\right].$$

Hereafter, unless otherwise stated, the PGMM family is taken to mean the family of 12 models discussed in this section (cf. Table 3.2).

3.5 Mixture of Common Factor Analyzers

3.5.1 The Model

Baek et al. (2010) argue that there may be situations where the mixture of factor analyzers model is not sufficiently parsimonious. They postulate that this might happen when p and/or G are not small. The same concern might also apply to other members of the PGMM family. To address this concern, Baek et al. (2010) build on the work of Yoshida et al. (2004, 2006) to introduce a mixture of common factor analyzers (MCFA) model. This model assumes that $\mathbf{X}_i$ can be modelled as

$$\mathbf{X}_i = \mathbf{\Lambda}\mathbf{U}_{ig} + \boldsymbol{\varepsilon}_{ig} \tag{3.16}$$

with probability π_g, for $i = 1, \ldots, n$ and $g = 1, \ldots, G$, where $\mathbf{\Lambda}$ is a $p \times q$ matrix of factor loadings, the $\mathbf{U}_{ig}$ are independently $\mathrm{N}(\boldsymbol{\xi}_g, \mathbf{\Omega}_g)$ and are independent of the $\boldsymbol{\varepsilon}_{ig}$, which are independently $\mathrm{N}(\mathbf{0}, \mathbf{\Psi})$, where $\mathbf{\Psi}$ is a $p \times p$ diagonal matrix with positive diagonal entries. Note that $\boldsymbol{\xi}_g$ is a q-dimensional vector and $\mathbf{\Omega}_g$ is a $q \times q$ covariance matrix. It follows that the density of $\mathbf{X}_i$ from the MCFA model is given by

$$f(\mathbf{x}_i \mid \vartheta) = \sum_{g=1}^{G} \pi_g \phi(\mathbf{x}_i \mid \mathbf{\Lambda}\boldsymbol{\xi}_g, \mathbf{\Lambda}\mathbf{\Omega}_g\mathbf{\Lambda}' + \mathbf{\Psi}), \tag{3.17}$$

where ϑ denotes the model parameters.

3.5.2 Parameter Estimation

Before detailing parameter estimation for the mixture of common factor analyzers model, it is helpful to note that

$$\left[\begin{array}{c} \mathbf{X}_i \\ \mathbf{U}_{ig} \end{array}\right]\bigg|_{z_{ig}=1} \sim \mathrm{N}\left(\left[\begin{array}{c} \mathbf{\Lambda}\boldsymbol{\xi}_g \\ \boldsymbol{\xi}_g \end{array}\right], \left[\begin{array}{cc} \mathbf{\Lambda}\mathbf{\Omega}_g\mathbf{\Lambda}' + \mathbf{\Psi} & \mathbf{\Lambda}\mathbf{\Omega}_g \\ \mathbf{\Omega}_g\mathbf{\Lambda}' & \mathbf{\Omega}_g \end{array}\right]\right).$$

It follows that

$$\mathbb{E}[\mathbf{U}_{ig} \mid \mathbf{x}_i, z_{ig} = 1] = \boldsymbol{\xi}_g + \boldsymbol{\beta}_g(\mathbf{x}_i - \mathbf{\Lambda}\boldsymbol{\xi}_g) \tag{3.18}$$

and

$$
\mathbb{E}[\mathbf{U}_{ig}\mathbf{U}'_{ig} \mid \mathbf{x}_i, z_{ig} = 1]
$$
$$
= \mathrm{Var}[\mathbf{U}_{ig} \mid \mathbf{x}_i, z_{ig} = 1] + \mathbb{E}[\mathbf{U}_{ig} \mid \mathbf{x}_i, z_{ig} = 1]\mathbb{E}[\mathbf{U}_{ig} \mid \mathbf{x}_i, z_{ig} = 1]' \quad (3.19)
$$
$$
= \mathbf{\Omega}_g - \boldsymbol{\beta}_g \mathbf{\Lambda}\mathbf{\Omega}_g + [\boldsymbol{\xi}_g + \boldsymbol{\beta}_g(\mathbf{x}_i - \mathbf{\Lambda}\boldsymbol{\xi}_g)][\boldsymbol{\xi}_g + \boldsymbol{\beta}_g(\mathbf{x}_i - \mathbf{\Lambda}\boldsymbol{\xi}_g)]',
$$

where $\boldsymbol{\beta}_g = \mathbf{\Omega}_g\mathbf{\Lambda}'(\mathbf{\Lambda}\mathbf{\Omega}_g\mathbf{\Lambda}' + \mathbf{\Psi})^{-1}$.

Baek et al. (2010) outline an EM algorithm for parameter estimation for the mixture of common factor analyzers. Again, denote by $\mathbf{z}_1, \ldots, \mathbf{z}_n$ the unobserved component labels, where $z_{ig} = 1$ if observation i belongs to component g and $z_{ig} = 0$ otherwise. The complete-data are taken to be the observed $\mathbf{x}_1, \ldots, \mathbf{x}_n$ together with the component membership labels z_{ig} and the latent factors $\mathbf{u}_{ig}$, for $i = 1, \ldots, n$ and $g = 1, \ldots, G$. The complete-data log-likelihood is given by

$$
l_c(\boldsymbol{\vartheta}) = \sum_{i=1}^{n}\sum_{g=1}^{G} z_{ig} \log\left[\pi_g \phi(\mathbf{x}_i \mid \mathbf{\Lambda}\mathbf{u}_{ig}, \mathbf{\Psi})\phi(\mathbf{u}_{ig} \mid \boldsymbol{\xi}_g, \mathbf{\Omega}_g)\right], \quad (3.20)
$$

and the expected values of the component membership labels are given by

$$
\hat{z}_{ig} = \frac{\hat{\pi}_g \phi(\mathbf{x}_i \mid \hat{\mathbf{\Lambda}}\hat{\boldsymbol{\xi}}_g, \hat{\mathbf{\Lambda}}\hat{\mathbf{\Omega}}_g\hat{\mathbf{\Lambda}}' + \hat{\mathbf{\Psi}})}{\sum_{h=1}^{G} \hat{\pi}_h f(\mathbf{x}_i \mid \hat{\mathbf{\Lambda}}\hat{\boldsymbol{\xi}}_h, \hat{\mathbf{\Lambda}}\hat{\mathbf{\Omega}}_h\hat{\mathbf{\Lambda}}' + \hat{\mathbf{\Psi}})}, \quad (3.21)
$$

for $i = 1, \ldots, n$ and $g = 1, \ldots, G$. The expected value of the complete-data log-likelihood can be written

$$
Q = \sum_{i=1}^{n}\sum_{g=1}^{G} \hat{z}_{ig} \left[\log \pi_g - p \log 2\pi - \frac{1}{2}\log|\mathbf{\Omega}_g| - \frac{1}{2}\log|\mathbf{\Psi}| \right.
$$
$$
- \frac{1}{2}\mathrm{tr}\left\{\mathbf{\Omega}_g^{-1}\mathbb{E}[(\mathbf{U}_{ig} - \boldsymbol{\xi}_g)(\mathbf{U}_{ig} - \boldsymbol{\xi}_g)' \mid \mathbf{x}_i]\right\}
$$
$$
\left. - \frac{1}{2}\mathrm{tr}\left\{\mathbf{\Psi}^{-1}\mathbb{E}[(\mathbf{X}_i - \mathbf{\Lambda}\mathbf{U}_{ig})(\mathbf{X}_i - \mathbf{\Lambda}\mathbf{U}_{ig})' \mid \mathbf{x}_i]\right\} \right].
$$

As usual, expectations are conditional on current parameter estimates. Maximizing Q with respect to π_g yields

$$
\hat{\pi}_g = \frac{n_g}{n}, \quad (3.22)
$$

where $n_g = \sum_{g=1}^{G} \hat{z}_{ig}$. Differentiating Q with respect to $\boldsymbol{\xi}_g$, $\mathbf{\Omega}_g^{-1}$, $\mathbf{\Lambda}$, and $\mathbf{\Psi}^{-1}$, respectively, gives the score functions

$$S_1(\boldsymbol{\vartheta}_{\mathrm{s}}) = \frac{\partial \mathcal{Q}}{\partial \boldsymbol{\xi}_g} = -\frac{1}{2}\boldsymbol{\Omega}_g^{-1}\sum_{i=1}^{n}\hat{z}_{ig}(\mathbb{E}[\mathbf{U}_{ig} \mid \mathbf{x}_i, z_{ig} = 1] - \boldsymbol{\xi}_g),$$

$$S_2(\boldsymbol{\vartheta}_{\mathrm{s}}) = \frac{\partial \mathcal{Q}}{\partial \boldsymbol{\Omega}_g^{-1}} = \frac{n_g}{2}\boldsymbol{\Omega}_g - \frac{1}{2}\sum_{i=1}^{n}\hat{z}_{ig}(\mathbb{V}\mathrm{ar}[\mathbf{U}_{ig} - \boldsymbol{\xi}_g \mid \mathbf{x}_i, z_{ig} = 1]$$
$$+ \mathbb{E}[\mathbf{U}_{ig} - \boldsymbol{\xi}_g \mid \mathbf{x}_i, z_{ig} = 1]\mathbb{E}[(\mathbf{U}_{ig} - \boldsymbol{\xi}_g)' \mid \mathbf{x}_i, z_{ig} = 1]),$$

$$S_3(\boldsymbol{\vartheta}_{\mathrm{s}}) = \frac{\partial \mathcal{Q}}{\partial \boldsymbol{\Lambda}}$$
$$= \sum_{i=1}^{n}\sum_{g=1}^{G}\hat{z}_{ig}\boldsymbol{\Psi}^{-1}(\mathbf{x}_i\mathbb{E}[\mathbf{U}_{ig}' \mid \mathbf{x}_i, z_{ig} = 1] - \boldsymbol{\Lambda}\mathbb{E}[\mathbf{U}_{ig}\mathbf{U}_{ig}' \mid \mathbf{x}_i, z_{ig} = 1]),$$

$$S_4(\boldsymbol{\vartheta}_{\mathrm{s}}) = \frac{\partial \mathcal{Q}}{\partial \boldsymbol{\Psi}^{-1}} = \frac{n}{2}\boldsymbol{\Psi} - \frac{1}{2}\sum_{i=1}^{n}\sum_{g=1}^{G}\hat{z}_{ig}(\mathbf{x}_i\mathbf{x}_i' - \boldsymbol{\Lambda}\mathbb{E}[\mathbf{U}_{ig} \mid \mathbf{x}_i, z_{ig} = 1]\mathbf{x}_i'$$
$$- \mathbf{x}_i\mathbb{E}[\mathbf{U}_{ig}' \mid \mathbf{x}_i, z_{ig} = 1]\boldsymbol{\Lambda}' + \boldsymbol{\Lambda}\mathbb{E}[\mathbf{U}_{ig}\mathbf{U}_{ig}' \mid \mathbf{x}_i, z_{ig} = 1]\boldsymbol{\Lambda}'),$$

where $\boldsymbol{\vartheta}_{\mathrm{s}} = (\boldsymbol{\xi}_g, \boldsymbol{\Omega}_g, \boldsymbol{\Lambda}, \boldsymbol{\Psi})$.

Solving $S_1(\hat{\boldsymbol{\xi}}_g^{\mathrm{new}}, \hat{\boldsymbol{\Omega}}_g^{\mathrm{new}}, \hat{\boldsymbol{\Lambda}}^{\mathrm{new}}, \hat{\boldsymbol{\Psi}}^{\mathrm{new}}) = \mathbf{0}$ gives

$$\hat{\boldsymbol{\xi}}_g^{\mathrm{new}} = \frac{1}{n_g}\sum_{i=1}^{n}\hat{z}_{ig}\mathbb{E}[\mathbf{U}_{ig} \mid \mathbf{x}_i, z_{ig} = 1] = \hat{\boldsymbol{\xi}}_g + \frac{1}{n_g}\sum_{i=1}^{n}\hat{z}_{ig}\hat{\boldsymbol{\beta}}_g(\mathbf{x}_i - \hat{\boldsymbol{\Lambda}}\hat{\boldsymbol{\xi}}_g),$$

and solving $S_2(\hat{\boldsymbol{\xi}}_g^{\mathrm{new}}, \hat{\boldsymbol{\Omega}}_g^{\mathrm{new}}, \hat{\boldsymbol{\Lambda}}^{\mathrm{new}}, \hat{\boldsymbol{\Psi}}^{\mathrm{new}}) = \mathbf{0}$ gives

$$\hat{\boldsymbol{\Omega}}_g^{\mathrm{new}} = (\mathbf{I}_q - \hat{\boldsymbol{\beta}}_g\hat{\boldsymbol{\Lambda}})\hat{\boldsymbol{\Omega}}_g + \frac{1}{n_g}\sum_{i=1}^{n}\hat{z}_{ig}\hat{\boldsymbol{\beta}}_g(\mathbf{x}_i - \hat{\boldsymbol{\Lambda}}\hat{\boldsymbol{\xi}}_g)(\mathbf{x}_i - \hat{\boldsymbol{\Lambda}}\hat{\boldsymbol{\xi}}_g)'\hat{\boldsymbol{\beta}}_g'.$$

Solving $S_3(\hat{\boldsymbol{\xi}}_g^{\mathrm{new}}, \hat{\boldsymbol{\Omega}}_g^{\mathrm{new}}, \hat{\boldsymbol{\Lambda}}^{\mathrm{new}}, \hat{\boldsymbol{\Psi}}^{\mathrm{new}}) = \mathbf{0}$ gives

$$\hat{\boldsymbol{\Lambda}}^{\mathrm{new}} = \left\{\sum_{i=1}^{n}\sum_{g=1}^{G}\hat{z}_{ig}\mathbf{x}_i[\hat{\boldsymbol{\xi}}_g + \hat{\boldsymbol{\beta}}_g(\mathbf{x}_i - \hat{\boldsymbol{\Lambda}}\hat{\boldsymbol{\xi}}_g)]'\right\}\left\{\sum_{i=1}^{n}\sum_{g=1}^{G}\hat{z}_{ig}(\mathbf{I}_q - \hat{\boldsymbol{\beta}}_g\hat{\boldsymbol{\Lambda}})\hat{\boldsymbol{\Omega}}_g\right.$$
$$\left. + \sum_{i=1}^{n}\sum_{g=1}^{G}\hat{z}_{ig}[\hat{\boldsymbol{\xi}}_g + \hat{\boldsymbol{\beta}}_g(\mathbf{x}_i - \hat{\boldsymbol{\Lambda}}\hat{\boldsymbol{\xi}}_g)][\hat{\boldsymbol{\xi}}_g + \hat{\boldsymbol{\beta}}_g(\mathbf{x}_i - \hat{\boldsymbol{\Lambda}}\hat{\boldsymbol{\xi}}_g)]'\right\}^{-1},$$

and solving $\mathrm{diag}\{S_4(\hat{\boldsymbol{\xi}}_g^{\mathrm{new}}, \hat{\boldsymbol{\Omega}}_g^{\mathrm{new}}, \hat{\boldsymbol{\Lambda}}^{\mathrm{new}}, \hat{\boldsymbol{\Psi}}^{\mathrm{new}})\} = \mathbf{0}$ gives

$$\hat{\boldsymbol{\Psi}}^{\mathrm{new}} = \frac{1}{n}\mathrm{diag}\left\{\sum_{i=1}^{n}\sum_{g=1}^{G}\hat{z}_{ig}\left[\mathbf{x}_i\mathbf{x}_i' - \hat{\boldsymbol{\Lambda}}\mathbb{E}[\mathbf{U}_{ig} \mid \mathbf{x}_i, z_{ig} = 1]\mathbf{x}_i'\right.\right.$$
$$\left.\left. - \mathbf{x}_i\mathbb{E}[\mathbf{U}_{ig}' \mid \mathbf{x}_i, z_{ig} = 1]\hat{\boldsymbol{\Lambda}}' + \hat{\boldsymbol{\Lambda}}\mathbb{E}[\mathbf{U}_{ig}\mathbf{U}_{ig}' \mid \mathbf{x}_i, z_{ig} = 1]\hat{\boldsymbol{\Lambda}}'\right]\right\}.$$

An EM Algorithm for the MCFA Model

An EM algorithm for the MCFA model can now be summarized.

EM Algorithm for MCFA

initialize $\hat{z}_{ig}$
initialize $\hat{\pi}_g, \hat{\boldsymbol{\xi}}_g, \hat{\boldsymbol{\Omega}}_g, \hat{\boldsymbol{\Lambda}}, \hat{\boldsymbol{\Psi}}$
while convergence criterion not met
    compute $\hat{\boldsymbol{\beta}}_g$
    update $\hat{\pi}_g, \hat{\boldsymbol{\xi}}_g^{\text{new}}, \hat{\boldsymbol{\Omega}}_g^{\text{new}}, \hat{\boldsymbol{\Lambda}}^{\text{new}}, \hat{\boldsymbol{\Psi}}^{\text{new}}$
    update $\hat{z}_{ig}$
    check convergence criterion
    $\hat{\boldsymbol{\xi}}_g \leftarrow \hat{\boldsymbol{\xi}}_g^{\text{new}}, \hat{\boldsymbol{\Omega}}_g \leftarrow \hat{\boldsymbol{\Omega}}_g^{\text{new}}, \hat{\boldsymbol{\Lambda}} \leftarrow \hat{\boldsymbol{\Lambda}}^{\text{new}}, \hat{\boldsymbol{\Psi}} \leftarrow \hat{\boldsymbol{\Psi}}^{\text{new}}$
end while

3.5.3 Discussion

The MCFA model places additional restrictions on the component means and covariance matrices compared to the mixture of factor analyzers model, thereby further reducing the number of parameters to be estimated. Consequently, the model is quite restrictive and, notably, is much more restrictive than the mixture of factor analyzers model. In fact, the MCFA model can be cast as a special case of the mixture of factor analyzers model (cf. Baek et al., 2010). Other than some situations in which the number of components G, the number of variables p, or both are very large, the mixture of factor analyzers model, or another member of the PGMM family, will almost certainly be preferable. One alternative to the MCFA model for larger values of p is discussed in Section 4.2. Another alternative is the heteroscedastic factor mixture analysis approach of Montanari and Viroli (2010).

3.6 Illustrations

3.6.1 x2 Data

First, consider the x2 dataset from the `mixture` package. This is a two-dimensional dataset generated from an EVE model with three components, $\pi_1 = \pi_2 = \pi_3 = 1/3$, and $n = 300$. As mentioned in Section 2.6.1, where they were analyzed using the GPCM models, the x2 data represent a very easy classification example and quite a straightforward clustering example. They are analyzed here only to illustrate that the PGMM models can perform very well on low-dimensional data. All 12 PGMMs are fitted to these data using the pgmmEM() function from **pgmm** for $G = 1, \ldots, 5$ components and $q = 1$ factor, using k-means starting values. The BIC selects a $G = 3$ component UUC model, i.e., a MPPCA model, with BIC $= -2001.45$. The corresponding MAP classifications correspond to the true values, save for the usual misclassification (cf. Section 2.6.1). Repeating this analysis with five random starts leads to selection of the same model.

3.6.2   Italian Wine Data

Next, consider the **wine** dataset from the **pgmm** package. The dataset contains 27 physical and chemical measurements (Table 3.3) on 178 wines grown in the same region in Italy but derived from three different cultivars: Barolo, Grignolino, and Barbera.

Table 3.3
Twenty-seven physical and chemical properties of the Italian wines.

Alcohol	Sugar-free extract	Fixed acidity
Tartaric acid	Malic acid	Uronic acids
pH	Ash	Alkalinity of ash
Potassium	Calcium	Magnesium
Phosphate	Chloride	Total phenols
Nonflavonoid phenols	Flavonoids	Proanthocyanins
OD_{280}/OD_{315} of diluted wines	Hue	Color intensity
OD_{280}/OD_{315} of flavonoids	Glycerol	2-3-butanediol
Total nitrogen	Proline	Methanol

This is the first illustration with data that are not low dimensional. These data are not particularly difficult as a classification example; within the literature, these data or a subset thereof have been used to illustrate clustering approaches. All 12 PGMMs are fitted to these data using the pgmmEM() function from **pgmm** for $G = 1, \ldots, 5$ and $q = 1, \ldots, 5$, using k-means starting values. The BIC (-11548.55) selects a CCUU model with $q = 5$ factors and $G = 4$ components. Because the number of factors is at the boundary of the range that was used, the analysis is repeated for $G = 1, \ldots, 5$ and $q = 1, \ldots, 8$. Now the BIC (-11479.09) selects a CUU model with $q = 6$ factors and $G = 3$ components. A cross-tabulation of the corresponding MAP classifications versus the true classes is given in Table 3.4, and only four samples are misclassified.

The analysis is repeated using five random starts with $G = 1, \ldots, 5$ and $q = 1, \ldots, 5$. The BIC selects a CUU model with $q = 4$ factors and $G = 3$ components. The BIC for this model is -11427.65, and the associated MAP classifications are very close to the true labels, with only two misclassifications (ARI $= 0.97$, Table 3.5).

To put the performance of the PGMM family of models in context, it is useful to consider the performance of the mixture of factor analyzers and MP-

Table 3.4
Cross-tabulation of the MAP classifications (A–C) associated with the selected PGMM, using k-means starts, against true classes for the `wine` data.

	A	B	C
Barolo	59	0	0
Grignolino	3	67	1
Barbera	0	0	48

Table 3.5
Cross-tabulation of the MAP classifications (A–C) associated with the selected PGMM, using random starts, against true classes for the `wine` data.

	A	B	C
Barolo	59	0	0
Grignolino	1	69	1
Barbera	0	0	48

PCA models on these data. In each case, k-means and five random starts are used for the same ranges of G and q as before. As usual, runs are repeated with a larger range in cases where the selected model is at the edge of the range. For the mixture of factor analyzers model, the BIC (-11710.18) selects $G = 2$ components and $q = 5$ factors; in effect, this model puts the Barolo and Grignolino wines into the same component (ARI = 0.47; Table 3.6). The selected MPPCA model has $G = 5$ components and $q = 5$ latent factors (BIC $= -11653.61$). Even with favourable merging of components, this MPPCA model still misclassifies 14 wines (Table 3.7).

Table 3.6
Cross-tabulation of the MAP classifications (A–B) associated with the selected mixture of factor analyzers model, chosen from amongst k-means and random starts, against true classes for the `wine` data.

	A	B
Barolo	59	0
Grignolino	67	4
Barbera	0	48

Table 3.7
Cross-tabulation of the MAP classifications (A–E) associated with the selected MPPCA model, chosen from amongst k-means and random starts, against true classes for the `wine` data.

	A	B	C	D	E
Barolo	59	0	0	0	0
Grignolino	5	51	6	9	0
Barbera	0	0	0	20	28

3.6.3   Italian Olive Oil Data

Forina and Tiscornia (1982) and Forina et al. (1983) report the percentage composition of eight fatty acids found by lipid fraction of 572 Italian olive oils, and these data are available as `olive` within `pgmm`. These data were used in Section 2.6.4, where they are discussed in detail. A classification scenario is simulated by taking every fourth olive oil to be unlabelled, with the effect that 25% of the oils are unlabelled. All PGMM models are run for $G = 9$ and $q = 1, \ldots, 6$, and the BIC (-5450.338) selects a UCU model with $q = 6$ factors (Table 3.8). The classification performance of this model is excellent, with ARI = 0.93. Furthermore, all points that are classified into the wrong area are still within the correct region, cf. the 4-2-3 block-diagonal structure of Table 3.8.

Table 3.8
Cross-tabulation of the MAP classifications (A–I) associated with the selected PGMM model against true classes for the `olive` data, where every fourth oil is treated as unlabelled.

	A	B	C	D	E	F	G	H	I
North Apulia	6	0	0	0	0	0	0	0	0
Calabria	0	14	0	0	0	0	0	0	0
South Apulia	0	0	49	2	0	0	0	0	0
Sicily	0	1	0	8	0	0	0	0	0
Inland Sardinia	0	0	0	0	16	0	0	0	0
Coastal Sardinia	0	0	0	0	0	9	0	0	0
East Liguria	0	0	0	0	0	0	12	0	0
West Liguria	0	0	0	0	0	0	1	12	0
Umbria	0	0	0	0	0	0	2	0	11

3.6.4 Alon Colon Cancer Data

Alon et al. (1999) present gene expression data for 62 colon tissue samples, of which 40 are tumours and the remaining 22 are normal. Affymetrix arrays are used to collect expression data for 6,500 genes. Alon et al. (1999) and McLachlan et al. (2002) focus their respective analyses on the 2,000 genes with the highest minimal intensity. McNicholas and Murphy (2010b) further reduce the number of genes from 2,000 to 461, and this subset is studied here.

First, consider clustering. Using the parallel computation approach described by McNicholas et al. (2010), McNicholas and Murphy (2010b) fitted the PGMM family to these data for $G = 2$ components and $q = 1, \ldots, 10$ factors, using ten different random starting values for the $\hat{z}_{ig}$. The BIC (-70937.72) selected a CUC model with $q = 6$ latent factors. The associated MAP classifications correspond very closely to tissue type (Table 3.9), with only five misclassifications and ARI $= 0.70$.

Table 3.9
Cross-tabulation of the MAP classifications (A,B) associated with the selected PGMM, using random starts, against tissue type for the Alon colon cancer data.

	A	B
Tumour	37	3
Normal	2	20

In this case, it is interesting to consider the PGMM model with the second best BIC value. This is a CCUU model with $q = 7$ latent factors and BIC $= -71063.71$ (cf. McNicholas and Murphy, 2010b). The associated MAP classifications do not correspond to tissue type and are similar to what McLachlan et al. (2002) call C_1 in their analysis. Getz et al. (2000) note that tissues 1–11 and 41–51 were extracted from the first 11 patients using a poly detector, while the remaining samples were taken from the other patients using total extraction of ribonucleic acid (RNA). The MAP classifications associated with the $q = 7$ factor CCUU model correspond closely to extraction method, with only eight misclassifications (Table 3.10; ARI $= 0.54$).

Because there are two alternative sets of labels, i.e., tissue type and extraction method, it is interesting to consider whether model-based classification is effective in each case. A classification scenario is simulated by taking every fourth tissue sample to be unlabelled. First, consider classification into tissue type. All PGMM models are run for $G = 2$ and $q = 1, \ldots, 10$, using the pgmm package, and the BIC (-70896.01) selects a CCC model with $q = 6$ factors (Table 3.11). The classification performance of this model is good, with three of the 15 unlabelled tissue samples misclassified.

Next consider classification into extraction method. All PGMM models are run for $G = 2$ and $q = 1, \ldots, 10$, and the BIC (-70196.04) selects a CUU

Table 3.10
Cross-tabulation of the MAP classifications (A,B) associated with the second-best PGMM, using random starts, against extraction method for the Alon colon cancer data.

	A	B
Poly detector	19	3
Total extraction of RNA	5	35

Table 3.11
Cross-tabulation of the MAP classifications (A,B) associated with the selected PGMM model against tissue type for the Alon colon cancer data, where every fourth tissue sample is treated as unlabelled.

	A	B
Tumour	8	2
Normal	1	4

Table 3.12
Cross-tabulation of the MAP classifications (A,B) associated with the selected PGMM model against extraction method for the Alon colon cancer data, where every fourth tissue sample is treated as unlabelled.

	A	B
Poly detector	4	0
Total extraction of RNA	2	9

model with $q = 8$ factors (Table 3.12). The classification performance of this model is good, with only two misclassifications.

Because these classification results are good but not excellent, it is interesting to repeat these analyses with a different set of unlabelled samples. Specifically, every fourth sample is taken as unlabelled starting at sample 5, i.e., tissue samples 5, 9, 13, etc. are taken as unlabelled. As before, all PGMM models are run for $G = 2$ and $q = 1, \ldots, 10$ in both cases. For classification into tissue type, the BIC (-71233.98) selects a CUC model with $q = 6$ factors. The classification performance of this model is excellent, with no misclassifications (Table 3.13).

Next, consider classification into extraction method. The BIC (-70480.51) selects a CUU model with $q = 8$ factors (Table 3.14). The classification performance of this model is also excellent, with no misclassifications.

Table 3.13
Cross-tabulation of the MAP classifications (A,B) associated with the selected PGMM model against tissue type for the Alon colon cancer data, where about one quarter of the tissue samples were treated as unlabelled.

	A	B
Tumour	10	0
Normal	0	5

Table 3.14
Cross-tabulation of the MAP classifications (A,B) associated with the selected PGMM model against extraction method for the Alon colon cancer data, where about one quarter of the tissue samples were treated as unlabelled.

	A	B
Poly detector	6	0
Total extraction of RNA	0	9

3.7 Comments

The mixture of factor analyzers and a family of mixture models based thereon, i.e., the PGMM family, can be used for clustering data with higher dimensions. However, model selection can become a problem when even a few hundred variables exist. Notably, in the cluster analysis of the Alon cancer data, McNicholas and Murphy (2010b) fix the number of components at $G = 2$ rather than running a range of values for G. The reason for this is that the BIC is not effective for estimating G in high-dimensional applications because the penalty term dominates. In such cases, more thought is needed on dimension reduction, e.g., via shrinkage (cf. Section 4.2). This point is interesting when one considers that mixtures of factor analyzers are often touted as methods for clustering high-dimensional data; however, this is really only the case if the number of clusters is known *a priori* and, even in such cases, the number of factors and model (i.e., covariance structure) cannot always be chosen reliably. An example in the next chapter will illustrate the BIC choosing the most simple model (CCC), with $q = 1$ factor, for $p = 250$ and $p = 500$, while also underestimating the value of G (Section 4.2, Table 4.1).

Looking at the examples in this chapter, one might get the impression that models with constrained factor loading matrices, i.e., $\Lambda_g = \Lambda$, are often selected by the BIC. This reflects the utility of borrowing information across components to estimate the covariances — constraining $\Lambda_g = \Lambda$ holds the covariances equal across components while allowing the variances to differ (unless the CCC or CCU model is being fitted). Experience suggests that, if only one PGMM model is going to be fitted to any but a low-dimensional dataset, the CUU model is a good choice.

Note that applying factor analysis, or principal components analysis, to data before clustering will generally not return results that are comparable to applying approaches such as those discussed in this chapter. This point has been made several times, including by Steinley and Brusco (2011b) who write that

> Although using data reduction techniques, such as factor analysis or principal components analysis, as a complementary technique can be helpful in terms of estimation, we caution users that the recovered clusters from the factor space often do not correspond to the cluster structure present in the manifest variables.

Bouveyron and Brunet–Saumard (2014) illustrate this point in an analysis of the `crabs` data from `MASS` library for R — these data are analyzed elsewhere herein, e.g., Section 4.7.2. Specifically, Bouveyron and Brunet–Saumard (2014) carry out a principal components analysis followed by model-based clustering using the GPCM family; they show that the classification results so obtained are far inferior to an analysis using the PGMM family on the (manifest) data. From this analysis, they quite eloquently conclude that

> ... when groups are localized in different subspaces, a global dimension reduction is not adapted for the clustering task. A efficient way to overcome such a problem is to use subspace clustering methods which model and cluster the data in group-specific low-dimensional subspaces.

Within the sphere of dimension reduction, the PGMM models and the MCFA model can be considered implicit dimension reduction techniques. In addition to dimension reduction, the MCFA approach also facilitates visualization. However, it is only recommended for consideration in situations with too many variables and/or components to use the mixture of factor analyzers model (Baek et al., 2010) and selection of the number of components and common factors remains a concern. Variable selection approaches are explicit dimension reduction techniques and are discussed as such in Chapter 4.

4

Dimension Reduction and
High-Dimensional Data

4.1 Implicit and Explicit Approaches

Suppose p-dimensional data vectors $\mathbf{x}_1, \ldots, \mathbf{x}_n$ are observed and p is large enough that dimension reduction is required. Note that the vagueness about how large p needs to be before it is considered "large" is intentional and necessary because the answer will depend on several factors, including the modelling process and the number of observations n. Note also that dimension reduction is often required, or at least helpful, even when p is not large because the presence of variables that are not helpful in discriminating groups can have a deleterious effect on clustering, or classification, performance.

Broadly, there are two ways to carry out dimension reduction: a subset of the p variables can be selected or the data can be mapped to a (much) lower-dimensional space. For reasons that will be apparent, the former approach can be referred to as explicit dimension reduction, whereas the latter is implicit. The mixture of factor analyzers model and extensions thereof are examples of implicit dimension reduction. However, as mentioned in Chapter 3, the MCFA approach is not recommended for general use and the PGMM family is not effective for (very) large values of p. The latter problem can be (partly) addressed by using a LASSO (least absolute shrinkage and selection operator)-penalized likelihood approach and model selection criterion (Bhattacharya and McNicholas, 2014), and work in this direction is discussed in Section 4.2. There are at least two other implicit dimension reduction techniques that deserve mention (GMMDR and HD-GMM) and these will be discussed in Sections 4.5 and 4.6. Similar to the latent factor-based approach, these methods carry out simultaneous clustering and dimension reduction. As Bouveyron and Brunet–Saumard (2014) point out in their excellent review, separating these two elements does not usually work if carried out sequentially; they give the particular example of clustering after principal component analysis (cf. Section 3.7). There are a number of explicit approaches by which variables are selected. The `clustvarsel` and `selvarclust` approaches have been used for the GPCM family of models (Section 4.4). More recently, the VSCC (variable selection for clustering and classification) approach has been developed and used in the same situation (Section 4.3).

4.2   PGMM Family in High-Dimensional Applications

As mentioned in Section 3.7, the cluster analysis of the Alon cancer data (Section 3.6.4) raises an interesting question: why fix $G = 2$ rather than run a range of values for G? The answer is that the BIC is not effective for estimating G in high-dimensional applications because the penalty term dominates. Recognizing this problem, Bhattacharya and McNicholas (2014) consider a LASSO-penalized likelihood approach and proceed to show that the LASSO-penalized BIC (LPBIC) can be used to effectively select the number of components in high dimensions, where the BIC fails. Specifically, they use the penalized log-likelihood

$$\log \mathcal{L}_{\text{pen}}(\boldsymbol{\vartheta}) = \log \left\{ \prod_{i=1}^{n} \sum_{g=1}^{G} \pi_g \phi(\mathbf{x}_i \mid \boldsymbol{\mu}_g, \boldsymbol{\Sigma}_g) \right\} - n\lambda_n \sum_{g=1}^{G} \pi_g \sum_{j=1}^{p} |\mu_{gj}|, \quad (4.1)$$

where μ_{gj} is the jth element in $\boldsymbol{\mu}_g$ and λ_n is a tuning parameter that depends on n. Denote

$$\varphi(\boldsymbol{\mu}) = \sum_{g=1}^{G} \pi_g \sum_{j=1}^{p} \varphi(\mu_{gj})$$

so that (4.1) can be written

$$\log \mathcal{L}_{\text{pen}}(\boldsymbol{\vartheta} \mid \mathbf{x}) = \log \mathcal{L}(\boldsymbol{\vartheta} \mid \mathbf{x}) - n\lambda_n \varphi(\boldsymbol{\mu}).$$

Following Heiser (1995) and others, Bhattacharya and McNicholas (2014) locally approximate the penalty using a quadratic function. Suppose $\hat{\boldsymbol{\mu}}$ is the estimate of $\boldsymbol{\mu}$ after a number of iterations; then their penalty can be locally approximated as

$$\varphi(\boldsymbol{\mu}) \approx n\lambda_n \sum_{g=1}^{G} \pi_g \sum_{j=1}^{p_g} \left[|\hat{\mu}_{gj}| + \frac{1}{2} \frac{\text{sign}\{\hat{\mu}_{gj}\}}{\hat{\mu}_{gj}} (\mu_{gj}^2 - \hat{\mu}_{gj}^2) \right], \quad (4.2)$$

where p_g is the number of non-zero elements in $\boldsymbol{\mu}_g$.

The approximation of the sum of the absolute values of the mean components in (4.2) is precisely the majorization of the absolute value of the mean components by quadratic majorization. De Leeuw (1994), Fan and Li (2001), and others use this majorization approach for maximizing the penalized likelihood. Moreover, Lange et al. (2000) prove the majorizing property of such functions in the context of various optimization transfer algorithms, including the EM algorithm. Van Deun et al. (2011) use this approach for maximizing the penalized log-likelihood in the context of a sparse principal components analysis approach for different penalty functions, including the LASSO.

Details on the derivation of the LPBIC and on parameter estimation from the associated penalized likelihood are given by Bhattacharya and McNicholas

(2014) and will not be repeated here. It is, however, useful to consider part of the simulation study performed therein. Data are generated from a $G = 3$ component, p-dimensional Gaussian mixture model for $p \in \{100, 250, 500\}$. The eight members of the PGMM family of McNicholas and Murphy (2008), and the LASSO-penalized analogues thereof, are fitted to these data for $G = 1, \ldots, 4$ and $q = 1, 2, 3$. The results (Table 4.1) show that the performance of the LPBIC does not seem to depend on the value of p, but the performance of the BIC clearly does.

Table 4.1
The models chosen by the LPBIC and BIC, respectively, for the high-dimensional data simulated by Bhattacharya and McNicholas (2014).

p	LPBIC				BIC			
	G	q	Model	ARI	G	q	Model	ARI
100	3	3	CUC	0.88	3	3	CUC	0.86
250	3	2	CUC	0.82	2	1	CCC	0.62
500	3	3	CUC	0.97	2	1	CCC	0.49

4.3 VSCC

The VSCC technique finds a subset of variables that simultaneously minimizes the within-group variance and maximizes the between-group variance, thereby resulting in variables that show separation between the desired groups. The within-group variance for variable j can be written

$$\mathcal{W}_j = \frac{\sum_{g=1}^{G} \sum_{i=1}^{n} z_{ig}(x_{ij} - \mu_{gj})^2}{n},$$

where x_{ij} is the value of variable j for observation i, μ_{gj} is the mean of variable j in component g, and n and z_{ig} have the usual meanings. The variance within variable j that is not accounted for by $\mathcal{W}_j$, i.e., $\sigma_j^2 - \mathcal{W}_j$, provides an indication of the variance between groups. In general, calculation of this residual variance is needed; however, if the data are standardized to have equal variance across variables, then any variable minimizing the within-group variance also maximizes the leftover variance. Accordingly, Andrews and McNicholas (2014) describe the VSCC method in terms of variables that are standardized to have zero mean and unit variance. The VSCC approach also uses the correlation between variables, which is denoted ρ_{jk} for variables j and k. If V represents the space of currently selected variables, then variable j is selected

if
$$|\rho_{jr}| < 1 - \mathcal{W}_j^m$$

for all $r \in V$, where $m \in \{1, \ldots, 5\}$ is fixed. VSCC is a step-wise approach that can be summarized as follows.

VSCC

calculate $\mathcal{W}_j$ for $j = 1, \ldots, p$

create **W**, a list containing the $\mathcal{W}_j$ sorted in ascending order

variable 1 (i.e., corresponding to W_1, the 1st entry in **W**) is placed into V

set $k = 1$

while $k < p$

    k++

    if $|\rho_{kr}| < 1 - W_k^m$ for all $r \in V$

        variable k is placed into V

    end if

end while

Andrews and McNicholas (2014) consider $m \in \{1, \ldots, 5\}$ and, from this point of view, there are five different VSCC algorithms. In general, the condition

$$|\rho_{kr}| < 1 - W_k^m \qquad (4.3)$$

in the VSCC algorithm could be replaced by a more general condition

$$|\rho_{kr}| < f(W_k),$$

where $f(W_k)$ is some suitable function. However, constructing such a suitable function is not a trivial matter. As Andrews and McNicholas (2014) point out, any relationship that will allow a variable with $\mathcal{W}_j = 1$ to be selected will not work, nor will any relationship that results in values outside $[0, 1]$. While there are other possibilities for constructing $f(W_k)$, Andrews and McNicholas (2014) consider (4.3) with $m \in \{1, \ldots, 5\}$, which results in up to five distinct subsets of variables (Table 4.2).

Table 4.2

The variance-correlation relationships considered by Andrews and McNicholas (2014).

Quintic	$|\rho_{kr}| < 1 - W_k^5$
Quartic	$|\rho_{kr}| < 1 - W_k^4$
Cubic	$|\rho_{kr}| < 1 - W_k^3$
Quadratic	$|\rho_{kr}| < 1 - W_k^2$
Linear	$|\rho_{kr}| < 1 - W_k$

When VSCC is used for clustering, the choice of subset must be made without specific knowledge of which one produces the best classifier. Andrews and McNicholas (2014) choose the subset that minimizes

$$\sum_{i=1}^{n}\sum_{g=1}^{G}\hat{z}_{ig} - \sum_{i=1}^{n}\max_{g}\{\hat{z}_{ig}\} = n - \sum_{i=1}^{n}\max_{g}\{\hat{z}_{ig}\},$$

which is equivalent to maximizing

$$\sum_{i=1}^{n}\max_{g}\{\hat{z}_{ig}\}. \qquad (4.4)$$

This criterion seems to work well (cf. Andrews and McNicholas, 2014); however, it only works for $G > 1$ because (4.4) always equals n when $G = 1$. When clustering with VSCC, the first step is to carry out an initial clustering using all of the variables, and the algorithm proceeds as follows.

VSCC for Clustering

perform initial clustering
carry out VSCC using the resulting clustering results
for each distinct variable subset (there will be up to five)
    apply model-based clustering
end for
return the variable subset and classifications for the best model

---

The initial clustering method can be any approach, model-based or otherwise. The `vscc` package uses the `mclust` software for both the initial clustering and clustering on the variable subset. While the `vscc` software can be useful, it is important to note that it is just one implementation of what is a very broadly applicable method. Furthermore, this implementation will be impractical in high dimensions because the number of free model parameters for some of the GPCM models is quadratic in data dimensionality. When using VSCC for model-based classification, a decision must be made regarding whether to use all observations or just the labelled observations to compute the $\mathcal{W}_j$. Otherwise, it proceeds in an analogous fashion to the clustering case outlined above. VSCC can also be used for model-based discriminant analysis (cf. Andrews and McNicholas, 2014).

4.4 clustvarsel and selvarclust

Raftery and Dean (2006) propose a variable selection method that utilizes a greedy search of the model space. Their approach is based on Bayes factors. Given data $\mathbf{x}$, the Bayes factor B_{12} for model $\mathcal{M}_1$ versus model $\mathcal{M}_2$ is

$$B_{12} = \frac{p(\mathbf{x} \mid \mathcal{M}_1)}{p(\mathbf{x} \mid \mathcal{M}_2)},$$

where

$$p(\mathbf{x} \mid \mathcal{M}_k) = \int p(\mathbf{x} \mid \boldsymbol{\theta}_k, \mathcal{M}_k) p(\boldsymbol{\theta}_k \mid \mathcal{M}_k) d\boldsymbol{\theta}_k,$$

$\boldsymbol{\theta}_k$ is the vector of parameters for model $\mathcal{M}_k$, and $p(\boldsymbol{\theta}_k \mid \mathcal{M}_k)$ is the prior distribution of $\mathcal{M}_k$ (Kass and Raftery, 1995). The approach of Raftery and Dean (2006) simultaneously selects a variable subset, the number of components, and the model, i.e., the GPCM covariance structure. This approach is implemented within the `clustvarsel` package (Dean et al., 2012) for R, and it can work well in some situations. However, because the number of free model parameters for some of the GPCM models is quadratic in data dimensionality, `clustvarsel` is largely ineffective in high dimensions.

A related approach is described by Maugis et al. (2009a,b) and implemented as the `selvarclust` software (Maugis, 2009), which is a command-line addition to the MIXMOD software (Biernacki et al., 2006). This approach relaxes the assumptions on the role of variables with the potential benefit of avoiding the over-penalization of independent variables. In general, the approach, as presented in Maugis et al. (2009b), involves three possible roles for variables: relevant variables, redundant variables, and independent variables.

4.5   GMMDR

Similar to `vscc`, `clustvarsel`, and `selvarclust`, the Gaussian mixture modelling and dimension reduction (GMMDR) approach (Scrucca, 2010) is based on the GPCM family of models, and builds on the sliced inverse regression work of Li (1991, 2000). The idea behind GMMDR is to find the smallest subspace that captures the clustering information contained within the data. To do this, GMMDR seeks those directions where the cluster (i.e., component) means $\boldsymbol{\mu}_g$ and the cluster covariances $\boldsymbol{\Sigma}_g$ vary the most, provided that each direction is $\boldsymbol{\Sigma}$-orthogonal to the others. These directions can be found via the the generalized eigen-decomposition of the kernel matrix

$$\mathbf{M}\mathbf{v}_i = l_i \boldsymbol{\Sigma} \mathbf{v}_i,$$

where $l_1 \geq l_2 \geq \cdots \geq l_d > 0$ and

$$\mathbf{v}_i' \boldsymbol{\Sigma} \mathbf{v}_j = \begin{cases} 1 & \text{if } i = j, \text{ and} \\ 0 & \text{otherwise} \end{cases}$$

(Scrucca, 2010). Note that there are $d \leq p$ directions that span the subspace. The kernel matrix contains the variations in cluster means

$$\mathbf{M}_{\mathrm{I}} = \sum_{g=1}^{G} \pi_g (\boldsymbol{\mu}_g - \boldsymbol{\mu})(\boldsymbol{\mu}_g - \boldsymbol{\mu})'$$

and the variations in cluster covariances

$$\mathbf{M}_{\mathrm{II}} = \sum_{g=1}^{G} \pi_g (\boldsymbol{\Sigma}_g - \bar{\boldsymbol{\Sigma}}) \boldsymbol{\Sigma}^{-1} (\boldsymbol{\Sigma}_g - \bar{\boldsymbol{\Sigma}})',$$

such that

$$\mathbf{M} = \mathbf{M}_{\mathrm{I}} \boldsymbol{\Sigma}^{-1} \mathbf{M}_{\mathrm{I}} + \mathbf{M}_{\mathrm{II}}.$$

Note that

$$\boldsymbol{\mu} = \sum_{g=1}^{G} \pi_g \boldsymbol{\mu}_g \quad \text{and} \quad \boldsymbol{\Sigma} = \frac{1}{n} \sum_{i=1}^{n} (\mathbf{x}_i - \boldsymbol{\mu})(\mathbf{x}_i - \boldsymbol{\mu})'$$

are the global mean and global covariance matrix, respectively, and

$$\bar{\boldsymbol{\Sigma}} = \sum_{g=1}^{G} \pi_g \boldsymbol{\Sigma}_g$$

is the pooled within-cluster covariance matrix.

The GMMDR directions are the eigenvectors $(\mathbf{v}_1, \ldots, \mathbf{v}_d) =: \boldsymbol{\beta}$. These eigenvectors, ordered according to the eigenvalues, form the basis of the dimension reduction subspace $\mathcal{S}(\boldsymbol{\beta})$. The projections of the mean and covariance onto $\mathcal{S}(\boldsymbol{\beta})$ are given by $\boldsymbol{\beta}' \boldsymbol{\mu}_g$ and $\boldsymbol{\beta}' \boldsymbol{\Sigma}_g \boldsymbol{\beta}$, respectively. The GMMDR variables are the projections of the $n \times p$ data matrix $(\mathbf{x}_1', \ldots, \mathbf{x}_n')'$ onto the subspace $\mathcal{S}(\boldsymbol{\beta})$ and can be computed as $(\mathbf{x}_1', \ldots, \mathbf{x}_n')' \boldsymbol{\beta}$. This estimation of GMMDR variables is a sort of feature extraction. Moreover, some of the estimated GMMDR variables may provide no clustering information and need to be removed. Scrucca (2010) removes them via a modified version of the variable selection method of Raftery and Dean (2006).

Scrucca (2014) extends the GMMDR approach to model-based discriminant analysis, and Morris and McNicholas (2016) apply GMMDR for model-based classification and model-based discriminant analysis. Morris et al. (2013) and Morris and McNicholas (2013, 2016) extend the GMMDR to mixtures of non-Gaussian distributions.

4.6 HD-GMM

Bouveyron et al. (2007a,b) introduce a family of 28 parsimonious, flexible

Gaussian models specifically designed for high-dimensional data. This family, called HD-GMM, can be applied for clustering or classification. The HD-GMM family is based on an eigen-decomposition of the component covariance matrices $\boldsymbol{\Sigma}_g$, which can be written

$$\boldsymbol{\Sigma}_g = \boldsymbol{\Gamma}_g \boldsymbol{\Delta}_g \boldsymbol{\Gamma}_g',$$

where $\boldsymbol{\Gamma}_g$ is a $p \times p$ orthogonal matrix of eigenvectors of $\boldsymbol{\Sigma}_g$ and $\boldsymbol{\Delta}_g$ is a $p \times p$ diagonal matrix containing the corresponding eigenvalues, in decreasing order. The idea behind the HD-GMM family is to reparametrize $\boldsymbol{\Delta}_g$ such that $\boldsymbol{\Sigma}_g$ has only $d_g + 1$ distinct eigenvalues. This is achieved via

$$\boldsymbol{\Delta}_g = \mathrm{diag}\{a_{g1}, \ldots, a_{gd_g}, b_g, \ldots, b_g\},$$

where the first $d_g < p$ values $a_{g1}, \ldots, a_{gd_g}$ represent the variance in the component-specific subspace and the other $p - d_g$ values b_g are the variance of the noise. The key assumption is that, conditional on the components, the noise variance for component g is isotropic and is within a subspace that is orthogonal to the subspace of the gth component. Although there are 28 HD-GMM models, the 16 with closed form estimators are often focused upon (Table 4.3).

Table 4.3
Nomenclature and the number of free covariance parameters for 16 members of the HD-GMM family.

Model	Number of Free Covariance Parameters
$[a_{gj}b_g\boldsymbol{\Gamma}_g d_g]$	$\sum_{g=1}^{G} d_g[p - (d_g + 1)/2] + \sum_{g=1}^{G} d_g + 2G$
$[a_{gj}b\boldsymbol{\Gamma}_g d_g]$	$\sum_{g=1}^{G} d_g[p - (d_g + 1)/2] + \sum_{g=1}^{G} d_g + 1 + G$
$[a_g b_g \boldsymbol{\Gamma}_g d_g]$	$\sum_{g=1}^{G} d_g[p - (d_g + 1)/2] + 3G$
$[ab_g\boldsymbol{\Gamma}_g d_g]$	$\sum_{g=1}^{G} d_g[p - (d_g + 1)/2] + 1 + 2G$
$[a_g b\boldsymbol{\Gamma}_g d_g]$	$\sum_{g=1}^{G} d_g[p - (d_g + 1)/2] + 1 + 2G$
$[ab\boldsymbol{\Gamma}_g d_g]$	$\sum_{g=1}^{G} d_g[p - (d_g + 1)/2] + 2 + G$
$[a_{gj}b_g\boldsymbol{\Gamma}_g d]$	$Gd[p - (d + 1)/2] + Gd + G + 1$
$[a_j b_g\boldsymbol{\Gamma}_g d]$	$Gd[p - (d + 1)/2] + d + G + 1$
$[a_{gj}b\boldsymbol{\Gamma}_g d]$	$Gd[p - (d + 1)/2] + Gd + 2$
$[a_j b\boldsymbol{\Gamma}_g d]$	$Gd[p - (d + 1)/2] + d + 2$
$[a_g b_g\boldsymbol{\Gamma}_g d]$	$Gd[p - (d + 1)/2] + 2G + 1$
$[ab_g\boldsymbol{\Gamma}_g d]$	$Gd[p - (d + 1)/2] + G + 2$
$[a_g b\boldsymbol{\Gamma}_g d]$	$Gd[p - (d + 1)/2] + G + 2$
$[ab\boldsymbol{\Gamma}_g d]$	$Gd[p - (d + 1)/2] + 3$
$[a_j b\boldsymbol{\Gamma} d]$	$d[p - (d + 1)/2] + d + 2$
$[ab\boldsymbol{\Gamma} d]$	$d[p - (d + 1)/2] + 3$

As Bouveyron and Brunet–Saumard (2014) point out, the HD-GMM family can be regarded as a generalization of the GPCM family or as a generalization of the MPPCA model. For instance, if $d_g = p - 1$, then the HD-GMM model $[a_{gj}b_g\Gamma_g d_g]$ is the same as the GPCM model VVV. Further, the HD-GMM model $[a_{gj}b_g\Gamma_g d]$ is equivalent to the MPPCA model. For the HD-GMM model $[a_g b_g\Gamma_g d_g]$, Bouveyron et al. (2011) show that the maximum likelihood estimate of the d_g is asymptotically consistent, a fact that has consequences for inference for isotropic PPCA (cf. Bouveyron et al., 2011).

4.7   Illustrations

4.7.1   Coffee Data

Streuli (1973) reports on the chemical composition of coffee samples collected from around the world. A total of 43 samples of the Arabica and Robusta species were collected from 29 countries, and 12 of the associated chemical constituents are available as the `coffee` data in `pgmm` (Table 4.4). A pairs plot of these data reveals that two variables, Fat and Caffeine, perfectly separate the Arabica and Robusta samples (Figure 4.1).

Table 4.4
The twelve chemical constituents given in the `coffee` data.

Water	Bean Weight	Extract Yield
pH Value	Free Acid	Mineral Content
Fat	Caffeine	Trigonelline
Chlorogenic Acid	Neochlorogenic Acid	Isochlorogenic Acid

Andrews and McNicholas (2014) conduct an analysis where the `vscc`, `clustvarsel`, and `selvarclust` packages are used to analyze the coffee data. The results (Table 4.5) indicate that `vscc` selects two variables (Fat and Caffeine) and gives perfect classification results. However, `clustvarsel` and `selvarclust` select six and seven variables, respectively, and the associated ARI values are much lower. Of course, *a posteriori* merging of components may help to mitigate this.

4.7.2   Leptograpsus Crabs

The `crabs` data are available in the `MASS` library for R. Therein are five morphometric measurements (Table 4.6) on two species of crabs (blue and orange), further separated into two genders. A pairs plot of the data reveal that the variables are highly correlated and this is a difficult clustering problem (Fig-

Figure 4.1
Pairs plot depicting the `coffee` data, where plotting symbol and shading reflect species, i.e., Arabica or Robusta.

Table 4.5
ARI values, number of components, and number of selected variables from `vscc`, `clustvarsel`, and `selvarclust` analyses of the coffee data.

	ARI	*G*	**No. Variables**
vscc (quadratic)	1.00	2	2
clustvarsel	0.41	3	6
selvarclust	0.37	4	7

ure 4.2). Raftery and Dean (2006) use these data to illustrate the efficacy of their variable selection approach.

Table 4.6
The five morphological measurements given in the `crabs` data, all measured in mm.

Frontal lobe size	Rear width
Carapace length	Carapace width
Body depth	

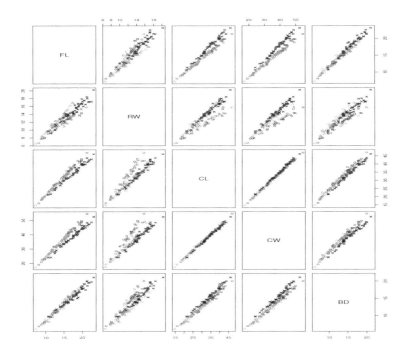

Figure 4.2
Pairs plot depicting the `crabs` data, where plotting symbol and shading reflect class, i.e., species and gender.

Andrews and McNicholas (2014) run `vscc`, `clustvarsel`, and `selvarclust` on these data. The results (Table 4.7) indicate that all approaches selected four of the five morphological measurements. Not surprisingly, one of the carapace measurements (carapace length) is dropped in each case (cf. Figure 4.2). The `vscc` and `clustvarsel` give the same classification performance (ARI = 0.76); however, despite selecting the same variables, `selvarclust` does not give the same classification performance. Andrews and McNicholas (2014) report that this reflects a difference in starting values used by `selvarclust`.

Table 4.7
ARI values, number of components, and number of selected variables from vscc, clustvarsel, and selvarclust analyses of the crabs data.

	ARI	G	**No. Variables**
vscc (quintic)	0.76	5	4
clustvarsel	0.76	5	4
selvarclust	0.50	5	4

4.7.3 Banknote Data

The banknote dataset from the mclust package was analyzed in Section 2.6.2. The dataset contains six measurements, all in mm, made on 100 genuine and 100 counterfeit Swiss 1,000-franc banknotes. As noted in Section 2.6.2, the variables Top and Diagonal provide very good discrimination (cf. Figure 2.7).

Andrews and McNicholas (2014) run vscc, clustvarsel, and selvarclust on these data. The results (Table 4.8) indicate that vscc selects four of the six measurements, while clustvarsel and selvarclust select five and three, respectively. Not surprisingly, all three approaches select Top and Diagonal. The vscc and clustvarsel approaches give results that are the same as the $G = 3$ and $G = 4$ component GPCM results, respectively, discussed in Section 2.6.2. The selvarclust approach gives inferior performance in terms of ARI; however, the situation can be improved by *a posteriori* merging.

Table 4.8
ARI values, number of components, and number of selected variables from vscc, clustvarsel, and selvarclust analyses of the banknote data.

	ARI	G	**No. Variables**
vscc (quadratic)	0.85	3	4
clustvarsel	0.67	4	5
selvarclust	0.25	8	3

4.7.4 Wisconsin Breast Cancer Data

Thus far, all illustrations have focused on explicit dimension reduction, i.e., variable selection. The Wisconsin breast cancer data can be used to illustrate implicit dimension reduction. The data are available from the UCI Machine Learning Repository and contain 30 quantitative features computed from digitized images of 569 fine needle aspirates of breast masses. Of the 569 samples, 357 are benign and 212 are malignant. Bouveyron and Brunet–Saumard (2014) analyze these data using the PGMM and HD-GMM families via the

`HDclassif` (Bergé et al., 2012, 2013) and `pgmm` packages, respectively. Note that they fix $G = 2$ components. The selected HD-GMM is the $[a_{gj}b_g\Gamma_g d]$ with $d = 11$, and the selected PGMM is a CUU model with $q = 2$ latent factors. As Bouveyron and Brunet–Saumard (2014) point out, the selected PGMM model is much more parsimonious but the classification performance of the selected HD-GMM is better (Table 4.9).

Table 4.9
Cross-tabulation of the classifications (A, B) associated with the selected PGMM and the selected HD-GMM models, respectively, for the Wisconsin breast cancer data.

	HD-GMM			PGMM	
	A	**B**		**A**	**B**
Benign	341	16		305	52
Malignant	40	172		67	145

4.7.5 Leukaemia Data

Golub et al. (1999) present gene expression microarray data on two forms of acute leukaemia: acute lymphoblastic leukaemia (ALL) and acute myeloid leukaemia (AML). Affymetrix arrays are used to obtain measurements for 7,129 genes on 72 tissues (25 AML and 47 ALL). McLachlan et al. (2002) reduce the dataset by removing genes with expression falling outside the interval $(100, 16000)$, and genes with expression satisfying max/min ≤ 5 or max $-$ min ≤ 500. In their analysis, McNicholas and Murphy (2010b) use the `select-genes` software (cf. McLachlan et al., 2002) to further reduce the number of genes to 2,030.

Bhattacharya and McNicholas (2014) analyze these 2,030 genes to compare the BIC and the LPBIC in the context of the PGMM family. Twenty different random starts for the initial $\hat{z}_{ig}$ are used and eight members of the PGMM family (Table 3.1) are run, in each case, for $G = 1, 2, 3$ components and $q = 1, \ldots, 6$ latent factors. The results (Table 4.10) reveal that the BIC chooses a CCU model with $G = 1$ component and $q = 2$ factors, whereas the LPBIC chooses a CUC model with $G = 2$ components and $q = 1$ factor. Not surprisingly, the classification perfromance of the model chosen using the LPBIC (ARI $= 0.47$) is greater than for the model chosen using the BIC — the BIC chooses a model with $G = 1$ component. The model selected using the LPBIC misclassifies 11 of the 72 samples (Table 4.11).

This example illustrates that the LPBIC can be effective in selecting the number of components for the PGMM family for a high-dimensional dataset ($p = 2,030$). Interestingly, the classification performance associated with the model with the second best LPBIC (CCU with $G = 2$ and $q = 2$) is better

Table 4.10
Comparison of the performance of the LPBIC and BIC for PGMM model selection for the leukaemia data.

	Value	Model	G	q	ARI
BIC	-400394	CCU	1	2	0
LPBIC	-391023	CUC	2	1	0.474

Table 4.11
Cross-tabulation of tissue type versus predicted classification for the PGMM model selected by the LPBIC (A, B) for the leukaemia data.

	A	B
ALL	39	8
AML	3	22

than that for the best model, with two fewer misclassifications. This underlines that, similar to the BIC, the model with the best LPBIC does not necessarily give the best clustering results. Of course, while this example serves as a good illustration of the LPBIC versus the BIC, it cannot be considered a thorough analysis of these data. Such an analysis would include, *inter alia*, many more starting values. In another illustration, McNicholas and Murphy (2010b) apply all 12 PGMMs to the same data, fixing $G = 2$, and the BIC selects a $q = 3$ factor CCUC model with an ARI $= 0.74$.

4.8   Comments

A host of implicit and explicit variable selection techniques is available for model-based clustering and classification. Some explicit approaches are discussed in this chapter. Various implicit approaches are discussed within this chapter and elsewhere herein. In general, explicit variable selection seems to be somewhat more difficult; for one, it becomes very difficult in higher dimensions. Implicit approaches seem to be more effective in higher dimensions; however, effective model selection remains a problem. Bouveyron and Brunet–Saumard (2014) provide an excellent review of work on model-based clustering of high-dimensional data.

Note that the suite of model-based approaches to dimension reduction and clustering, or classification, is growing quite rapidly and there is much work beyond that discussed in this chapter. Some recent examples of interesting alternatives include the work of Celeux et al. (2014) and Marbac and Sedki (2015). Finally, although their focus is not specifically on model-based clustering, Steinley and Brusco (2008) provide an interesting empirical comparison of several variable selection procedures for clustering.

5

Mixtures of Distributions with Varying Tail Weight

5.1 Mixtures of Multivariate t-Distributions

The first, and perhaps most natural, departure from the Gaussian mixture model is the mixture of multivariate t-distributions. McLachlan and Peel (1998) motivate the multivariate t-distribution as a heavy-tailed alternative to the multivariate Gaussian distribution by first considering the Gaussian scale mixture model

$$(1 - \epsilon)\phi(\mathbf{x} \mid \boldsymbol{\mu}, \boldsymbol{\Sigma}) + \epsilon\phi(\mathbf{x} \mid \boldsymbol{\mu}, c\boldsymbol{\Sigma}), \tag{5.1}$$

where c is large, ϵ is small, and $\phi(\mathbf{x} \mid \boldsymbol{\mu}, \boldsymbol{\Sigma})$ is the density of a multivariate Gaussian distribution with mean $\boldsymbol{\mu}$ and covariance matrix $\boldsymbol{\Sigma}$. Now (5.1) can be written

$$\int \phi(\mathbf{x} \mid \boldsymbol{\mu}, \boldsymbol{\Sigma}/w)dH(w), \tag{5.2}$$

where H is the probability distribution placing mass $(1 - \epsilon)$ at the point $w = 1$ and mass ϵ at the point $w = 1/c$. Then the multivariate t-distribution is obtained when H in (5.2) is replaced by the probability density of a random variable $W \sim \text{gamma}(\nu/2, \nu/2)$, where ν denotes degrees of freedom (McLachlan and Peel, 1998). Note that the gamma(a, b) density is

$$h(w \mid \alpha, \beta) = \frac{\beta^\alpha w^{\alpha-1} \exp\{-\beta w\}}{\Gamma(\alpha)},$$

for $w > 0$ with parameters $\alpha, \beta \in \mathbb{R}^+$. The density for the multivariate t-distribution is

$$f_t(\mathbf{x} \mid \boldsymbol{\mu}, \boldsymbol{\Sigma}, \nu) = \frac{\Gamma\left([\nu + p]/2\right)|\boldsymbol{\Sigma}|^{-1/2}}{(\pi\nu)^{p/2}\Gamma\left(\nu/2\right)\left[1 + \delta(\mathbf{x}, \boldsymbol{\mu} \mid \boldsymbol{\Sigma})/\nu\right]^{(\nu+p)/2}}, \tag{5.3}$$

with mean $\boldsymbol{\mu}$, scale matrix $\boldsymbol{\Sigma}$, and degrees of freedom ν, and where

$$\delta(\mathbf{x}, \boldsymbol{\mu} \mid \boldsymbol{\Sigma}) = (\mathbf{x} - \boldsymbol{\mu})'\boldsymbol{\Sigma}^{-1}(\mathbf{x} - \boldsymbol{\mu}) \tag{5.4}$$

is the squared Mahalanobis distance between $\mathbf{x}$ and $\boldsymbol{\mu}$. The density for a mixture of multivariate t-distributions is given by

$$f(\mathbf{x} \mid \boldsymbol{\vartheta}) = \sum_{g=1}^{G} \pi_g f_t(\mathbf{x} \mid \boldsymbol{\mu}_g, \boldsymbol{\Sigma}_g, \nu_g). \qquad (5.5)$$

The mixture of multivariate t-distributions model (5.5) has only G more free parameters than the mixture of multivariate Gaussian distributions. Parameter estimation proceeds in a similar fashion except for the presence of the w_{ig} and the need to estimate the degrees of freedom. Andrews and McNicholas (2012) introduce a t-analogue of the GPCM family of models by imposing the constraints in Table 2.1 on the component scale matrices $\boldsymbol{\Sigma}_g$. Andrews and McNicholas (2012) follow Andrews and McNicholas (2011a) by also considering the constraint $\nu_g = \nu$, which can lead to improved classification performance. Strictly, Andrews and McNicholas (2012) only consider analogues of 12 of the 14 GPCMs; however, all 14 are considered in the `teigen` package for R. Herein the associated family of 28 mixtures of multivariate t-distributions models is called the tEIGEN family (see Table 5.9).

Andrews and McNicholas (2012) use a multicycle ECM algorithm for parameter estimation for members of the tEIGEN family. The complete-data comprise the observed $\mathbf{x}_i$, the characteristic weights w_{ig}, and the component memberships z_{ig}, for $i = 1, \ldots, n$ and $g = 1, \ldots, G$. The complete-data log-likelihood for the most general tEIGEN model is

$$l_c(\boldsymbol{\vartheta}) = \sum_{g=1}^{G} \sum_{i=1}^{n} z_{ig} \log \left[\pi_g \phi(\mathbf{x}_i \mid \boldsymbol{\mu}_g, \boldsymbol{\Sigma}_g / w_{ig}) h \left(w_{ig} \mid \nu_g/2, \nu_g/2 \right) \right],$$

where $W_{ig} \sim \text{gamma}(\nu_g/2, \nu_g/2)$. At each E-step, the expected value of the component indicator variables, i.e.,

$$\hat{z}_{ig} = \frac{\hat{\pi}_g f_t(\mathbf{x} \mid \hat{\boldsymbol{\mu}}_g, \hat{\boldsymbol{\Sigma}}_g, \hat{\nu}_g)}{\sum_{h=1}^{G} \hat{\pi}_h f_t(\mathbf{x} \mid \hat{\boldsymbol{\mu}}_h, \hat{\boldsymbol{\Sigma}}_h, \hat{\nu}_h)},$$

and the expected values of the weights, i.e.,

$$\hat{w}_{ig} = \frac{\hat{\nu}_g + p}{\hat{\nu}_g + \delta(\mathbf{x}_i, \hat{\boldsymbol{\mu}}_g \mid \hat{\boldsymbol{\Sigma}}_g)}$$

are updated for $i = 1, \ldots, n$ and $g = 1, \ldots, G$. As usual, expectations are conditional on the current parameter estimates.

In the first CM-step, the mixing proportions and component means are updated using

$$\hat{\pi}_g = \frac{n_g}{n} \quad \text{and} \quad \hat{\boldsymbol{\mu}}_g = \frac{\sum_{i=1}^{n} \hat{z}_{ig} \hat{w}_{ig} \mathbf{x}_i}{\sum_{i=1}^{n} \hat{z}_{ig} \hat{w}_{ig}},$$

respectively, where $n_g = \sum_{i=1}^{n} \hat{z}_{ig}$. The degrees of freedom are also updated. When no constraint is imposed, the update for $\hat{\nu}_g$ is given by solving

$$1 - \varphi\left(\frac{\hat{\nu}_g^{\text{new}}}{2}\right) + \frac{1}{n_g} \sum_{i=1}^{n} \hat{z}_{ig}(\log \hat{w}_{ig} - \hat{w}_{ig}) + \log\left(\frac{\hat{\nu}_g^{\text{new}}}{2}\right)$$
$$+ \varphi\left(\frac{\hat{\nu}_g + p}{2}\right) - \log\left(\frac{\hat{\nu}_g + p}{2}\right) = 0, \tag{5.6}$$

for $\hat{\nu}_g^{\text{new}}$, where $\varphi(\cdot)$ is the digamma function. In the case where the constraint $\nu_g = \nu$ is used, the update is given by solving

$$1 - \varphi\left(\frac{\hat{\nu}^{\text{new}}}{2}\right) + \frac{1}{n} \sum_{g=1}^{G} \sum_{i=1}^{n} \hat{z}_{ig}(\log \hat{w}_{ig} - \hat{w}_{ig}) + \log\left(\frac{\hat{\nu}^{\text{new}}}{2}\right)$$
$$+ \varphi\left(\frac{\hat{\nu} + p}{2}\right) - \log\left(\frac{\hat{\nu} + p}{2}\right) = 0,$$

for $\hat{\nu}^{\text{new}}$.

In the second CM-step, the component scale matrices are updated. These updates are similar to their Gaussian analogues (cf. Celeux and Govaert, 1995; Browne and McNicholas, 2014a). For illustrative purposes, consider the member of the *t*EIGEN family with $\boldsymbol{\Gamma}_g = \boldsymbol{\Gamma}$ and $\boldsymbol{\Delta}_g = \boldsymbol{\Delta}$. The complete-data log-likelihood is

$$l_c = C + \sum_{i=1}^{n} \sum_{g=1}^{G} z_{ig} p \log \lambda_g$$
$$- \sum_{i=1}^{n} \sum_{g=1}^{G} \frac{z_{ig} w_{ig}}{\lambda_g} \text{tr}\{(\mathbf{x}_i - \boldsymbol{\mu}_g)(\mathbf{x}_i - \boldsymbol{\mu}_g)'(\boldsymbol{\Gamma}\boldsymbol{\Delta}\boldsymbol{\Gamma}')^{-1}\}, \tag{5.7}$$

where C is constant with respect to λ_g, $\boldsymbol{\Gamma}$, and $\boldsymbol{\Delta}$. Setting $\boldsymbol{\Gamma}\boldsymbol{\Delta}\boldsymbol{\Gamma}' = \boldsymbol{\Xi}$ and taking the conditional expected value of (5.7) gives

$$Q = C + p \sum_{g=1}^{G} n_g \log \lambda_g + \sum_{g=1}^{G} \frac{n_g}{\lambda_g} \text{tr}\{\mathbf{S}_g \boldsymbol{\Xi}^{-1}\}.$$

Differentiating Q with respect to λ_g and setting the result equal to zero gives

$$\hat{\lambda}_g = \frac{1}{p} \text{tr}\{\mathbf{S}_g \boldsymbol{\Xi}^{-1}\},$$

where

$$\mathbf{S}_g = \frac{1}{n_g} \sum_{i=1}^{n} \hat{z}_{ig} \hat{w}_{ig} (\mathbf{x}_i - \hat{\boldsymbol{\mu}}_g)(\mathbf{x}_i - \hat{\boldsymbol{\mu}}_g)'$$

can be interpreted as a weighted sample covariance matrix. Utilizing Theorem A.1 from Celeux and Govaert (1995), the update for $\boldsymbol{\Xi}$ is

$$\hat{\boldsymbol{\Xi}} = \frac{(1/\hat{\lambda}_g) \sum_{g=1}^{G} \mathbf{S}_g}{\left| (1/\hat{\lambda}_g) \sum_{g=1}^{G} \mathbf{S}_g \right|^{1/p}},$$

leading to $\hat{\boldsymbol{\Gamma}}$ as the matrix of eigenvectors of $\hat{\boldsymbol{\Xi}}$ and $\hat{\boldsymbol{\Delta}}$ as a diagonal matrix of the eigenvalues of $\hat{\boldsymbol{\Xi}}$.

Of course, for the *t*EIGEN models with fully unconstrained scale matrices, i.e., UUUC and UUUU (see Table 5.9), the update for the scale can be written

$$\hat{\boldsymbol{\Sigma}}_g = \mathbf{S}_g.$$

Similarly, for the fully constrained scale matrices, i.e., CCCC and CCCU (see Table 5.9), the update is

$$\hat{\boldsymbol{\Sigma}} = \sum_{g=1}^{G} \pi_g \mathbf{S}_g.$$

These four *t*EIGEN models appeared in the model-based classification literature in their own right, i.e., prior to the introduction of the *t*EIGEN family (cf. Andrews et al., 2011).

For illustration, consider the EM algorithm for model-based clustering using the entirely unconstrained *t*EIGEN model, i.e., UUUU.

Multicycle ECM Algorithm for Clustering via *t*-Mixtures

initialize $\hat{w}_{ig}$ and $\hat{z}_{ig}$
while convergence criterion not met
    update $\hat{\pi}_g = n_g/n$
    update $\hat{\boldsymbol{\mu}}_g = \sum_{i=1}^{n} \hat{z}_{ig} \hat{w}_{ig} \mathbf{x}_i / \sum_{i=1}^{n} \hat{z}_{ig} \hat{w}_{ig}$
    update $\hat{\nu}_g$ by solving for $\hat{\nu}_g^{\text{new}}$ in (5.6)
    update $\hat{z}_{ig}$ and $\hat{w}_{ig}$
    update $\hat{\boldsymbol{\Sigma}}_g = \mathbf{S}_g$
    update $\hat{z}_{ig}$ and $\hat{w}_{ig}$
    check convergence criterion
end while

5.2   Mixtures of Power Exponential Distributions

While mixtures of multivariate *t*-distributions have been the most popular approach for clustering with heavier tail weight, the multivariate power exponential (MPE) distribution has emerged as an alternative and is used for clustering by Dang et al. (2015). A random vector $\mathbf{X}$ follows a p-dimensional MPE distribution (Landsman and Valdez, 2003) if the density is of the form

$$h(\mathbf{x} \mid \boldsymbol{\mu}, \boldsymbol{\Sigma}, r, s) = c_p |\boldsymbol{\Sigma}|^{-1/2} \exp \left\{ -\frac{r}{2^s} \delta(\mathbf{x}, \boldsymbol{\mu} \mid \boldsymbol{\Sigma})^s \right\}, \qquad (5.8)$$

where

$$c_p = \frac{s \Gamma \left(\frac{p}{2} \right)}{(2\pi)^{p/2} \Gamma \left(\frac{p}{2s} \right)} r^{p/(2s)},$$

$\boldsymbol{\mu}$ is the location (and mean), $\boldsymbol{\Sigma}$ is the scale matrix, and $r, s > 0$ control the shape. Note that $\delta(\mathbf{x}, \boldsymbol{\mu} \mid \boldsymbol{\Sigma})$ has the usual meaning, cf. (5.4). The MPE is an elliptical distribution and a multivariate Kotz-type distribution; however, the formulation in (5.8) is not identifiable because the density with $\vartheta = \{\boldsymbol{\mu}, \boldsymbol{\Sigma}^*, r^*, s\}$, where $\boldsymbol{\Sigma}^* = \boldsymbol{\Sigma}/2$ and $r^* = r/2^s$, is the same as the density in (5.8).

Dang et al. (2015) use the parametrization given by Gómez et al. (1998), where the density of a random vector $\mathbf{X}$ following a p-dimensional MPE distribution is

$$f_{\mathrm{PE}}(\mathbf{x} \mid \boldsymbol{\mu}, \boldsymbol{\Sigma}, \beta) = k |\boldsymbol{\Sigma}|^{-1/2} \exp \left\{ -\frac{1}{2} \delta(\mathbf{x}, \boldsymbol{\mu} \mid \boldsymbol{\Sigma})^\beta \right\}, \qquad (5.9)$$

where

$$k = \frac{p \Gamma \left(\frac{p}{2} \right)}{\pi^{p/2} \Gamma \left(1 + \frac{p}{2\beta} \right) 2^{1+p/(2\beta)}},$$

$\boldsymbol{\mu}$ is the location (also the mean), $\boldsymbol{\Sigma}$ is the scale matrix, and β determines the kurtosis. Note that (5.9) follows from (5.8) by setting $r = 2^{\beta-1}$ and $s = \beta$.

As pointed out by Gómez et al. (1998), the covariance matrix and multidimensional kurtosis coefficient for the MPE distribution are

$$\mathbb{V}\mathrm{ar}[\mathbf{X}] = \frac{2^{1/\beta} \Gamma \left(\frac{p+2}{2\beta} \right)}{p \Gamma \left(\frac{p}{2\beta} \right)} \boldsymbol{\Sigma}$$

and

$$\gamma_2(\mathbf{X}) = \mathbb{E} \left\{ \left[(\mathbf{X} - \boldsymbol{\mu})' \mathbb{V}\mathrm{ar}[\mathbf{X}]^{-1} (\mathbf{X} - \boldsymbol{\mu}) \right]^2 \right\} - p(p+2)$$

$$= \frac{p^2 \Gamma \left(\frac{p}{2\beta} \right) \Gamma \left(\frac{p+4}{2\beta} \right)}{\Gamma^2 \left(\frac{p+2}{2\beta} \right)} - p(p+2),$$

respectively (Mardia et al., 1980; Gómez et al., 1998).

Depending on the value of β, two kinds of distributions can be obtained. For $0 < \beta < 1$, a leptokurtic distribution is obtained, which is characterized by a thinner peak and heavier tails compared to the Gaussian distribution. In this case, i.e., $\beta \in (0, 1)$, the MPE distribution is a scale mixture of

Gaussian distributions (Gómez–Sánchez–Manzano et al., 2008). For $\beta > 1$, a platykurtic distribution is obtained, which is characterized by a flatter peak and thinner tails compared to the Gaussian distribution. Some well-known distributions arise as special or limiting cases of the MPE distribution, e.g., a double-exponential distribution ($\beta = 0.5$), a Gaussian distribution ($\beta = 1$), and a multivariate uniform distribution ($\beta \to \infty$). Some of the shapes that can be accommodated using the MPE distribution are given in Figure 5.1. For comparison, a similar plot is given for the multivariate t-distribution in Figure 5.2. While the plots in Figure 5.1 are visibly different, differences between the plots in Figure 5.2 are much more difficult to ascertain. For this reason, Figure 5.3 is included and gives the contour plots corresponding to the density plots in Figure 5.2. This simple visual comparison highlights the relatively mild effect of varying ν in the multivariate t-distribution compared to varying β in the MPE distribution. In fact, the density shapes shown for the multivariate t-distribution (Figure 5.2) look very similar to the MPE distribution for β close to 1; however, for other values of β the MPE distribution has no comparable multivariate t-distribution.

Some examples of applications of the MPE distribution are given by Lindsey (1999), Cho and Bui (2005), and Verdoolaege et al. (2008). Dang et al. (2015) point out that the MPE distribution has been underused because of difficulties in estimating the covariance over the entire support of the shape parameter $\beta \in (0, \infty)$. Parameter estimation is difficult for the MPE distribution because none of the parameter estimates are available in closed form. To simplify parameter estimation, some work has focused on the special case where $\beta \in (0, 1)$, e.g., Gómez–Sánchez–Manzano et al. (2008), Bombrun et al. (2012), and Pascal et al. (2013). Zhang and Liang (2010) use a mixture of MPE distributions for clustering, focusing on the special case where $\beta \in (0, 2)$ — as Dang et al. (2015) point out, the fixed-point algorithm that Zhang and Liang (2010) use is guaranteed to converge for $\beta \in (0, 2)$ but will diverge for $\beta \geq 2$.

Dang et al. (2015) use a generalized EM (GEM) algorithm (Dempster et al., 1977) that is effective for $0 < \beta < \infty$. This approach utilizes MM algorithms (Hunter and Lange, 2000b) as well as accelerated line searches on the orthogonal Stiefel manifold (Absil et al., 2008; Browne and McNicholas, 2014c). A mixture of MPE distributions is given by

$$f(\mathbf{x} \mid \boldsymbol{\vartheta}) = \sum_{g=1}^{G} \pi_g f_{\text{PE}}\left(\mathbf{x} \mid \boldsymbol{\mu}_g, \boldsymbol{\Sigma}_g, \beta_g\right),$$

where $\boldsymbol{\vartheta}$ denotes all parameters. Zhang and Liang (2010) show that mixtures of MPE distributions are identifiable. Recognizing the need for parsimony in the component scale matrices, Dang et al. (2015) use analogues of some members of the GPCM family, which, along with the option to constrain $\beta_g = \beta$, leads to a family of 16 mixture models called the eigen-decomposed power exponential mixtures (ePEM) family (see Table 5.10).

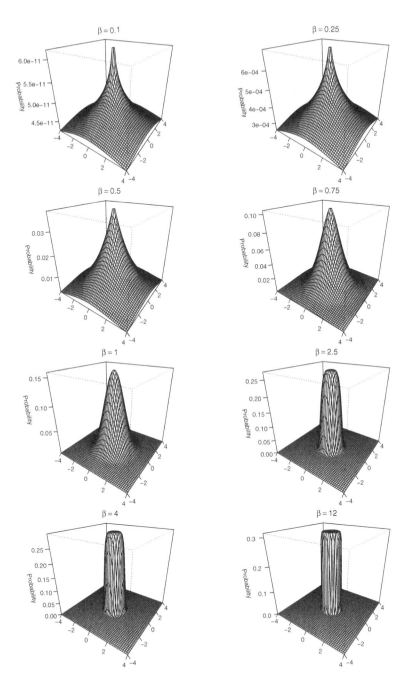

Figure 5.1
Density plots for the MPE distribution ($p = 2$) for different values of β.

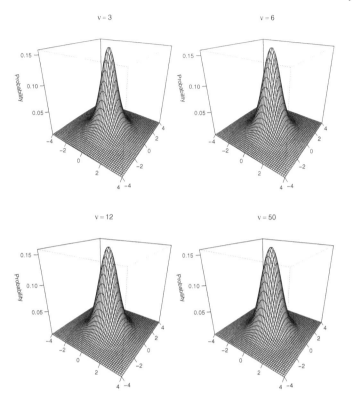

Figure 5.2
Density plots for the multivariate t-distribution $(p = 2)$ for different values of degrees of freedom.

The likelihood for $\mathbf{x}_1, \dots, \mathbf{x}_n$ from the MPE mixture model is

$$\mathcal{L}(\boldsymbol{\vartheta}) = \prod_{i=1}^{n} \sum_{g=1}^{G} \pi_g k_g |\boldsymbol{\Sigma}_g|^{-1/2} \exp\left\{ -\frac{1}{2} \delta(\mathbf{x}_i, \boldsymbol{\mu}_g \mid \boldsymbol{\Sigma}_g)^{\beta_g} \right\},$$

where

$$k_g = \frac{p\,\Gamma\,(p/2)}{\pi^{p/2}\Gamma\,(1 + p/2\beta_g)\, 2^{1+p/(2\beta_g)}}.$$

Defining z_{ig} in the usual way, the complete-data log-likelihood is given by

$$l_c(\boldsymbol{\vartheta}) = \sum_{i=1}^{n} \sum_{g=1}^{G} z_{ig} \log \left[\pi_g k_g |\boldsymbol{\Sigma}_g|^{-1/2} \exp\left\{ -\frac{\delta(\mathbf{x}_i, \boldsymbol{\mu}_g \mid \boldsymbol{\Sigma}_g)^{\beta_g}}{2} \right\} \right].$$

To compute the expected complete-data log-likelihood, $\mathcal{Q}$, we need the (con-

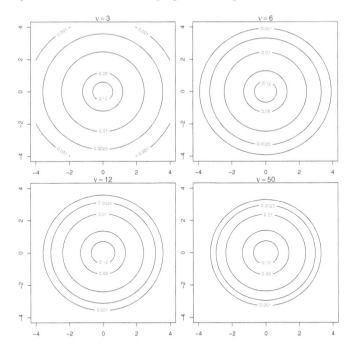

Figure 5.3
Contour plots for the multivariate t-distribution ($p = 2$) for different values of degrees of freedom.

ditional) expected values

$$\hat{z}_{ig} = \frac{\hat{\pi}_g f_{\mathrm{PE}}(\mathbf{x}_i \mid \hat{\boldsymbol{\mu}}_g, \hat{\boldsymbol{\Sigma}}_g, \hat{\beta}_g)}{\sum_{h=1}^{G} \hat{\pi}_h f_{\mathrm{PE}}(\mathbf{x}_i \mid \hat{\boldsymbol{\mu}}_h, \hat{\boldsymbol{\Sigma}}_h, \hat{\beta}_h)}, \tag{5.10}$$

for $i = 1, \ldots, n$ and $g = 1, \ldots, G$.

The M-step update for $\hat{\pi}_g$ is, as usual,

$$\hat{\pi}_g = \frac{n_g}{n},$$

where $n_g = \sum_{i=1}^{n} \hat{z}_{ig}$. Updates for $\hat{\boldsymbol{\mu}}_g$, $\hat{\beta}_g$, and the constituent elements of $\hat{\boldsymbol{\Sigma}}_g$ are not available in closed form. A Newton–Raphson update is used for $\hat{\boldsymbol{\mu}}_g$ (cf. Dang et al., 2015). An update for $\hat{\beta}_g$ can be obtained by solving the equation

$$\frac{pn_g}{\hat{\beta}_g^2} \varphi \left(1 + \frac{p}{2\hat{\beta}_g} \right) + \frac{pn_g \log 2}{\hat{\beta}_g^2}$$

$$- \sum_{i=1}^{n} \hat{z}_{ig} [\log \delta(\mathbf{x}_i, \boldsymbol{\mu}_g \mid \boldsymbol{\Sigma}_g)] \delta(\mathbf{x}_i, \boldsymbol{\mu}_g \mid \boldsymbol{\Sigma}_g)^{\hat{\beta}_g} = 0$$

for $\hat{\beta}_g$, where $\varphi(\cdot)$ is the digamma function. Alternatively, a Newton–Raphson method can be implemented (cf. Dang et al., 2015).

The scale updates will depend on the model (see Table 5.10) and, for several models, there are two solutions depending on the current estimate of $\hat{\beta}_g$. Consider the model with no constraints, i.e., VVVV. The expected value of the complete-data log-likelihood can be written

$$Q(\Sigma_g) = C + \sum_{i=1}^{n} \sum_{g=1}^{G} \frac{\hat{z}_{ig}}{2} \left\{ \log |\Sigma_g|^{-1} - \delta(\mathbf{x}_i, \boldsymbol{\mu}_g \mid \Sigma_g)^{\beta_g} \right\},$$

where C is constant with respect to Σ_g.

First, consider the case $\hat{\beta}_g \in (0, 1)$. A minorization-maximization framework — one type of MM algorithm — is used for estimation. This approach exploits the fact that

$$\operatorname{tr} \left\{ \Sigma_g^{-1} (\mathbf{x}_i - \hat{\boldsymbol{\mu}}_g)(\mathbf{x}_i - \hat{\boldsymbol{\mu}}_g)' \right\}^{\hat{\beta}_g}$$

is concave for $\hat{\beta}_g \in (0, 1)$. A surrogate function is then constructed using the supporting hyperplane inequality and further details are given by Dang et al. (2015). Now consider the case $\hat{\beta}_g \in [1, \infty)$. The component precision matrix can be decomposed as

$$\Sigma_g^{-1} = \Gamma_g \Delta_g^{-1} \Gamma_g',$$

where Γ_g is an orthonormal matrix and Δ_g is a diagonal matrix of eigenvalues. Dang et al. (2015) write

$$\Sigma_g^{-1} = \Gamma_g \mathbf{K}_g^{1/\hat{\beta}_g} \Gamma_g',$$

where $\mathbf{K}_g^{-1/\hat{\beta}_g} = \Delta_g$, and let $\mathbf{v}_{ig} = \Gamma_g'(\mathbf{x}_i - \boldsymbol{\mu}_g)$. Noting that, for $i = 1, \ldots, n$, the function

$$\operatorname{tr} \left\{ (\mathbf{x}_i - \boldsymbol{\mu}_g)' \hat{\Gamma}_g \mathbf{K}_g^{1/\hat{\beta}_g} \hat{\Gamma}_g' (\mathbf{x}_i - \boldsymbol{\mu}_g) \right\}^{\hat{\beta}_g} = \operatorname{tr} \left\{ \mathbf{v}_{ig}' \mathbf{K}_g^{1/\hat{\beta}_g} \mathbf{v}_{ig} \right\}^{\hat{\beta}_g}$$

$$= \left[\sum_{j=1}^{p} \kappa_{gj}^{1/\hat{\beta}_g} v_{igj}^2 \right]^{\hat{\beta}_g}$$

is concave with respect to $\operatorname{diag}\{\mathbf{K}_g\}$, a surrogate function can be constructed. An update for $\hat{\Delta}_g$ follows quite naturally, and an accelerated line search on the orthogonal Stiefel manifold (cf. Browne and McNicholas, 2014c) is used to update $\hat{\Gamma}_g$ — extensive details are available in Dang et al. (2015).

For illustration, consider the GEM algorithm for model-based clustering using the entirely unconstrained ePEM model, i.e., VVVV.

GEM Algorithm for Clustering via MPE Mixtures

initialize $\hat{z}_{ig}$, $\hat{\beta}_g$, and $\hat{\boldsymbol{\mu}}_g$

initialize $\hat{\boldsymbol{\Sigma}}_g$

update $\hat{z}_{ig}$

while convergence criterion not met

    update $\hat{\pi}_g$, $\hat{\beta}_g$, $\hat{\boldsymbol{\mu}}_g$, and $\hat{\boldsymbol{\Sigma}}_g$

    update $\hat{z}_{ig}$

    check convergence criterion

end while

5.3 Illustrations

5.3.1 Overview

The members of the ePEM, *t*EIGEN, and GPCM families are fitted to five real datasets: the x2 data from `mixture`, the `body` data from `gclus`, the `diabetes` data from `mclust`, the `f.voles` data from `MASS`, and the `crabs` data from `MASS`. To facilitate a direct comparison, only the $\boldsymbol{\Sigma}_g$ decompositions available for the ePEM family are used for the *t*EIGEN and GPCM families. In each case, all models are fitted for $G = 1, \ldots, 5$ and the BIC is used for model selection.

5.3.2 x2 Data

As seen in previous applications of Gaussian mixtures, the x2 data are easy to cluster. Not surprisingly, the selected model in all three cases has $G = 3$ components and misclassifies only the usual point (cf. Section 2.6.1). The selected ePEM model is a VVVE model with $\hat{\beta} = 1.09$ so that the components are effectively Gaussian. The selected *t*EIGEN model is also very close to a Gaussian mixture, i.e., a VVVE model with $\hat{\nu} = 50.05$. The selected GPCM is a VVE model — note that the EVE model is not available within the subset of models run.

5.3.3 Body Data

Next, consider the `body` data from `gclus`. Heinz et al. (2003) report data on 21 body dimension measurements as well as weight, height, age, and gender for 247 men and 260 women. The 507 people involved in this study were active individuals, i.e., exercising several hours a week, and most were in their twenties and thirties. The selected ePEM model is a $G = 2$ component EEEV model and each component has heavy tails, i.e., $\hat{\boldsymbol{\beta}} = (0.57, 0.56)'$. The associated classification performance is very good (ARI = 0.94; Table 5.1). The selected *t*EIGEN model is a $G = 3$ component EEEE model and each component has $\hat{\nu} = 23.43$ degrees of freedom. The associated classification performance is good (ARI = 0.80; Table 5.2) but not quite as good as for the ePEM models. The selected GPCM model is a $G = 3$ component EEE model, and the associated classification performance is similar to the selected *t*EIGEN model (Table 5.3).

Table 5.1
Cross-tabulation of the MAP classifications (A, B) associated with the selected ePEM against true classes for the body data.

	A	B
Female	256	4
Male	4	243

Table 5.2
Cross-tabulation of the MAP classifications (A–C) associated with the selected *t*EIGEN against true classes for the body data.

	A	B	C
Female	257	0	3
Male	5	47	195

Table 5.3
Cross-tabulation of the MAP classifications (A–C) associated with the selected GPCM against true classes for the body data.

	A	B	C
Female	257	0	3
Male	5	45	197

5.3.4   Diabetes Data

Reaven and Miller (1979) study the relationship between chemical and overt diabetes in 145 non-obese adults and the associated data are available as diabetes within mclust. For each subject, three measurements are taken: the degree of glucose intolerance, insulin response to oral glucose, and insulin resistance. As Reaven and Miller (1979) point out, the associated 3D plot (see Figure 5.4) is a "boomerang with two wings and a fat middle". Reaven and Miller (1979) proceed to classify the patients into three groups (normal, chemical diabetes, overt diabetes) on the basis of their oral glucose tolerance. The "wings" represent patients with chemical and overt diabetes, respectively, and this is best seen on a pairs plot (Figure 5.5).

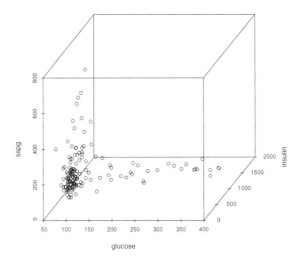

Figure 5.4
A 3D plot for the `diabetes` data.

For the analyses here, the selected models from all three families yield similar classification results, each with a total of 20 misclassifications. The selected ePEM model has $\hat{\beta} = 1.07$ in each component, suggesting components that are close to Gaussian, and the associated classifications are given in Table 5.4. The selected `teigen` model also has relatively high (50.30) degrees of freedom in each component, implying component shapes that are close to Gaussian.

Table 5.4
Cross-tabulation of the MAP classifications (A–C) associated with the selected ePEM against true classes for the `diabetes` data.

	A	**B**	**C**
Normal	26	9	1
Overt Diabetes	4	72	0
Chemical Diabetes	6	0	27

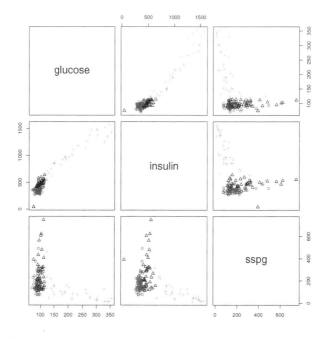

Figure 5.5
A pairs plot for the `diabetes` data, where shading and plotting symbol represents class, i.e., normal, chemical, and overt.

5.3.5   Female Voles Data

The `f.voles` data are available within the `Flury` package. These data contain age and six morphometric measurements for 86 female voles from two species: *Microtus californicus* and *M. ochrogaster*. They were previously discussed and analyzed in Section 2.6.3. For the analyses here, the selected models from all three families yield the same classification results (Table 5.5).

Table 5.5
Cross-tabulation of the MAP classifications (A, B) associated with the selected ePEM against true classes for the `f.voles` data.

	A	B
M. californicus	41	0
M. ochrogaster	2	43

5.3.6 Leptograpsus Crabs Data

The crabs data from MASS were previously considered in Section 4.7.2. There are five length measurements on two different colour forms of the crabs, further separated into two genders. The selected ePEM model is a $G = 3$ component VVVE model and each component is a platykurtic distribution, i.e., the peak is quite flat and the tails are thin, with $\hat{\beta} = 50.05$. The MAP classifications separate the male and female blue crabs quite well but put all of the orange crabs in the same component (Table 5.6; ARI = 0.60). The selected tEIGEN model is a $G = 4$ component EEVV model with $\hat{\nu} = (18.88, 120.31, 162.68, 6.23)$ degrees of freedom. The associated classification performance is better than for the ePEM model (Table 5.7; ARI = 0.81). The selected GPCM model is a $G = 3$ component VVV model, and the associated classifications put the male crabs in one component and try to separate the female crabs by colour (Table 5.8; ARI = 0.61).

Table 5.6
Cross-tabulation of the MAP classifications (A–C) associated with the selected ePEM against true classes for the crabs data.

	A	B	C
Blue Male	39	11	0
Blue Female	0	50	0
Orange Male	0	0	50
Orange Female	0	0	50

Table 5.7
Cross-tabulation of the MAP classifications (A–D) associated with the selected tEIGEN model against true classes for the crabs data.

	A	B	C	D
Blue Male	39	11	0	0
Blue Female	0	49	0	1
Orange Male	0	0	50	0
Orange Female	0	0	4	46

Table 5.8
Cross-tabulation of the MAP classifications (A–C) associated with the selected GPCM against true classes for the `crabs` data.

	A	B	C
Blue Male	38	12	0
Blue Female	0	49	1
Orange Male	50	0	0
Orange Female	2	0	48

5.4 Comments

The mixture of multivariate t-distributions and the MPE model accommodate tail weight in different ways. As illustrated via bivariate density plots (Figures 5.1 and 5.2), the MPE distribution allows a more notable departure from the multivariate Gaussian distribution than the multivariate t-distribution. Put another way, the multivariate t-distribution is a relatively gentle departure from the Gaussian distribution, with tails gradually becoming heavier as ν decreases; however, the MPE distribution moves away from the Gaussian rather quickly as β moves away from 1 (in either direction).

The results of the analyses in Section 5.3 are summarized in Table 5.11. The performance of the tEIGEN and GPCM families must be taken in context with the fact that only members with analogues in the ePEM family were run. However, doing it this way permits a direct comparison and, of course, the larger GPCM and tEIGEN families may not do any better. For three of the datasets, all three approaches gave identical or almost identical classification results. For the `body` data, the ePEM family outperformed the other two families and, for the `crabs` data, the tEIGEN gave the best classification performance. This suggests that there is room for all three approaches. Notably, the fact that the ePEM family allows much more flexibility does not mean that the mixture of MPE distributions will necessarily outperform the mixture of multivariate t-distributions in a given situation. For example, the mixture of multivariate t-distributions may be preferable in situations where clusters contain outliers, and further investigation in this direction is warranted. See Section 9.4 for further discussion of robust clustering. Finally, it is noteworthy that the mixture of factor analyzers model, other members of the PGMM family, and the MCFA model have been extended to the mixture of multivariate t-distributions (cf. McLachlan et al., 2007; Andrews and McNicholas, 2011a,b; Baek and McLachlan, 2011; McNicholas and Subedi, 2012; Steane et al., 2012; Lin et al., 2014).

Table 5.9

Nomenclature and number of covariance parameters for each member of the tEIGEN family, with constrained (C), unconstrained (U), and identity (I) elements.

Model	$\lambda_g = \lambda$	$\Gamma_g = \Gamma$	$\Delta_g = \Delta$	$\nu_g = \nu$	No. Free Covariance Parameters
CIIC	C	—	—	C	$1+1$
CIIU	C	—	—	U	$1+G$
UIIC	U	—	—	C	$(G-1)+1$
UIIU	U	—	—	U	$(G-1)+G$
CICC	C	—	C	C	$p+1$
CICU	C	—	C	U	$p+G$
UICC	U	—	C	C	$p+(G-1)+1$
UICU	U	—	C	U	$p+(G-1)+G$
CIUC	C	—	U	C	$Gp-(G-1)+1$
CIUU	C	—	U	U	$Gp-(G-1)+G$
UIUC	U	—	U	C	$Gp+1$
UIUU	U	—	U	U	$Gp+G$
CCCC	C	C	C	C	$[p(p+1)/2]+1$
CCCU	C	C	C	U	$[p(p+1)/2]+G$
UCCC	U	C	C	C	$[p(p+1)/2]+(G-1)+1$
UCCU	U	C	C	U	$[p(p+1)/2]+(G-1)+G$
CUCC	C	U	C	C	$G[p(p+1)/2]-(G-1)(p)+1$
CUCU	C	U	C	U	$G[p(p+1)/2]-(G-1)(p)+G$
UUCC	U	U	C	C	$G[p(p+1)/2]-(G-1)(p-1)+1$
UUCU	U	U	C	U	$G[p(p+1)/2]-(G-1)(p-1)+G$
CCUC	C	C	U	C	$[p(p+1)/2]-(G-1)(p-1)+1$
CCUU	C	C	U	U	$[p(p+1)/2]+(G-1)(p-1)+G$
CUUC	C	U	U	C	$G[p(p+1)/2]-(G-1)+1$
CUUU	C	U	U	U	$G[p(p+1)/2]-(G-1)+G$
UCUC	U	C	U	C	$G[p(p+1)/2]+(G-1)(p)+1$
UCUU	U	C	U	U	$G[p(p+1)/2]+(G-1)(p)+G$
UUUC	U	U	U	C	$G[p(p+1)/2]+1$
UUUU	U	U	U	U	$G[p(p+1)/2]+G$

Table 5.10

Nomenclature and number of covariance parameters for each member of the ePEM family, with values of certain parameters being equal (E) or variable (V) across components or equal to the identity matrix (I).

Model	λ_g	$\mathbf{\Gamma}_g$	$\mathbf{\Delta}_g$	β_g	No. Free Scale Parameters
EIIE	E	I	I	E	1
EIIV	E	I	I	V	1
VIIE	V	I	I	E	G
VIIV	V	I	I	V	G
EEIE	E	E	I	E	p
EEIV	E	E	I	V	p
VVIE	V	V	I	E	Gp
VVIV	V	V	I	V	Gp
EEEE	E	E	E	E	$p(p+1)/2$
EEEV	E	E	E	V	$p(p+1)/2$
EEVE	E	E	V	E	$Gp(p+1)/2 - (G-1)p$
EEVV	E	E	V	V	$Gp(p+1)/2 - (G-1)p$
VVEE	V	V	E	E	$p(p+1)/2 + (G-1)p$
VVEV	V	V	E	V	$p(p+1)/2 + (G-1)p$
VVVE	V	V	V	E	$Gp(p+1)/2$
VVVV	V	V	V	V	$Gp(p+1)/2$

Table 5.11

ARI values for the selected models from running ePEM, as well as the analogous models from teigen and mixture, on five datasets.

Data	n	p	Classes	ePEM	teigen	mixture
x2	300	2	3	0.99 ($G = 3$; VVVE)	0.99 ($G = 3$; VVVE)	0.99 ($G = 3$;VVE)
body	507	24	2	0.94 ($G = 2$; EEEV)	0.80 ($G = 3$; EEEE)	0.80 ($G = 3$; EEE)
diabetes	145	3	3	0.66 ($G = 3$; VVVE)	0.67 ($G = 3$; VVVE)	0.66 ($G = 3$; VVV)
f.voles	86	7	2	0.91 ($G = 2$; EEEV)	0.91 ($G = 2$; EEEE)	0.91 ($G = 2$; EEE)
crabs	200	5	4	0.60 ($G = 3$; VVVE)	0.81 ($G = 4$; EEVV)	0.61 ($G = 3$; VVV)

6

Mixtures of Generalized Hyperbolic Distributions

6.1 Overview

Until a few years after the turn of the century, almost all work on clustering and classification using mixture models had been based on Gaussian mixture models (e.g., Banfield and Raftery, 1993; Celeux and Govaert, 1995; Ghahramani and Hinton, 1997; Tipping and Bishop, 1999; McLachlan and Peel, 2000b; Fraley and Raftery, 2002b). Furthermore, what little work there was on non-Gaussian mixtures was on mixtures of multivariate t-distributions (e.g., McLachlan and Peel, 1998; Peel and McLachlan, 2000). A little beyond the turn of the century, work on t-mixtures burgeoned into a substantial subfield of mixture model-based classification (e.g., McLachlan et al., 2007; Andrews et al., 2011; Andrews and McNicholas, 2011a,b, 2012; Baek and McLachlan, 2011; McNicholas and Subedi, 2012; Steane et al., 2012; Lin et al., 2014). Around the same time, work on mixtures of skewed distributions took off, including work on skew-normal mixtures (e.g., Lin, 2009; Lin et al., 2016), skew-t mixtures (e.g., Lin, 2010; Lee and McLachlan, 2011, 2014; Vrbik and McNicholas, 2012, 2014; Murray et al., 2014a), shifted asymmetric Laplace mixtures (Franczak et al., 2014), variance-gamma mixtures (McNicholas et al., 2014), multivariate normal-inverse Gaussian mixtures (Karlis and Santourian, 2009; Subedi and McNicholas, 2014; O'Hagan et al., 2016), and generalized hyperbolic mixtures (Browne and McNicholas, 2015).

6.2 Generalized Inverse Gaussian Distribution

6.2.1 A Parameterization

The density of a random variable W following a generalized inverse Gaussian (GIG) distribution can be written

$$q(w \mid a, b, \lambda) = \frac{(a/b)^{\lambda/2} w^{\lambda-1}}{2K_\lambda(\sqrt{ab})} \exp\left\{ -\frac{aw + b/w}{2} \right\}, \tag{6.1}$$

for $w > 0$, where $a, b \in \mathbb{R}^+$, $\lambda \in \mathbb{R}$, and K_λ is the modified Bessel function of the third kind with index λ. There are several well-known special cases of the GIG distribution, including the gamma distribution ($b = 0$, $\lambda > 0$) and the inverse Gaussian distribution ($\lambda = -1/2$). The GIG distribution has been extensively studied within the literature, including work by Barndorff–Nielsen and Halgreen (1977), Blæsild (1978), Halgreen (1979), and Jørgensen (1982). It has some attractive properties, including the tractability of the following expected values:

$$\mathbb{E}\left[W\right] = \sqrt{\frac{b}{a}} \frac{K_{\lambda+1}\left(\sqrt{ab}\right)}{K_\lambda(\sqrt{ab})}, \tag{6.2}$$

$$\mathbb{E}\left[1/W\right] = \sqrt{\frac{a}{b}} \frac{K_{\lambda+1}\left(\sqrt{ab}\right)}{K_\lambda(\sqrt{ab})} - \frac{2\lambda}{b}, \tag{6.3}$$

$$\mathbb{E}[\log W] = \log\left(\sqrt{\frac{b}{a}}\right) + \frac{1}{K_\lambda(\sqrt{ab})}\frac{\partial}{\partial\lambda}K_\lambda(\sqrt{ab}). \tag{6.4}$$

Because there are other parameterizations of the GIG distributions, it is helpful to write $W \sim \text{GIG}(a, b, \lambda)$ to indicate that a random variable W has the GIG density as parameterized in (6.1).

6.2.2   An Alternative Parameterization

Consider the parameterization of the GIG used in (6.1), and write $\omega = \sqrt{ab}$ and $\eta = \sqrt{b/a}$. This results in a different, and sometimes more meaningful, parameterization of the GIG density,

$$h(w \mid \omega, \eta, \lambda) = \frac{(w/\eta)^{\lambda-1}}{2\eta K_\lambda(\omega)} \exp\left\{-\frac{\omega}{2}\left(\frac{w}{\eta} + \frac{\eta}{w}\right)\right\}, \tag{6.5}$$

where $\eta > 0$ is a scale parameter, $\omega > 0$ is a concentration parameter, and λ is an index parameter. To avoid confusion, write $W \sim \text{I}(\omega, \eta, \lambda)$ to indicate that a random variable W has the GIG density as parameterized in (6.5). The tractability of the following expected values is, as before, useful:

$$\mathbb{E}\left[W\right] = \eta\frac{K_{\lambda+1}\left(\omega\right)}{K_\lambda\left(\omega\right)},$$

$$\mathbb{E}\left[1/W\right] = \frac{1}{\eta}\frac{K_{\lambda-1}\left(\omega\right)}{K_\lambda\left(\omega\right)} = \frac{1}{\eta}\frac{K_{\lambda+1}\left(\omega\right)}{K_\lambda\left(\omega\right)} - \frac{2\lambda}{\omega\eta},$$

$$\mathbb{E}\left[\log W\right] = \log\eta + \frac{1}{K_\lambda\left(\omega\right)}\frac{\partial}{\partial v}K_\lambda\left(\omega\right).$$

6.3 Mixtures of Shifted Asymmetric Laplace Distributions

6.3.1 Shifted Asymmetric Laplace Distribution

The density of a p-dimensional random vector $\mathbf{V}$ from a centralized asymmetric Laplace (CAL) distribution (Kotz et al., 2001) has density

$$f(\mathbf{v} \mid \boldsymbol{\alpha}, \boldsymbol{\Sigma}) = \frac{2 \exp\{\mathbf{v}'\boldsymbol{\Sigma}^{-1}\boldsymbol{\alpha}\}}{(2\pi)^{p/2}|\boldsymbol{\Sigma}|^{1/2}} \left(\frac{\mathbf{v}'\boldsymbol{\Sigma}^{-1}\mathbf{v}}{2 + \boldsymbol{\alpha}'\boldsymbol{\Sigma}^{-1}\boldsymbol{\alpha}} \right)^{\lambda/2} K_\lambda(u), \qquad (6.6)$$

where $\lambda = (2 - p)/2$, $u = \sqrt{(2 + \boldsymbol{\alpha}'\boldsymbol{\Sigma}^{-1}\boldsymbol{\alpha})(\mathbf{v}'\boldsymbol{\Sigma}^{-1}\mathbf{v})}$, $\boldsymbol{\Sigma}$ is a scale matrix, and $\boldsymbol{\alpha} \in \mathbb{R}^p$ is the skewness parameter. Following Kotz et al. (2001), use the notation $\mathbf{V} \sim \mathrm{AL}_p(\boldsymbol{\alpha}, \boldsymbol{\Sigma})$ to indicate that the random variable $\mathbf{V}$ follows a p-dimensional CAL distribution. A mixture of CAL distributions would not be effective for clustering because each component would have the same location, i.e., the origin. To address this problem, Franczak et al. (2014) consider a random vector $\mathbf{X} = (\mathbf{V} + \boldsymbol{\mu}) \sim \mathrm{SAL}_p(\boldsymbol{\mu}, \boldsymbol{\Sigma}, \boldsymbol{\alpha})$, where $\boldsymbol{\mu} \in \mathbb{R}^p$ is a location parameter and $\mathrm{SAL}_p(\boldsymbol{\mu}, \boldsymbol{\Sigma}, \boldsymbol{\alpha})$ denotes a p-dimensional shifted asymmetric Laplace (SAL) distribution with density

$$f_{\mathrm{SAL}}(\mathbf{x} \mid \boldsymbol{\mu}, \boldsymbol{\Sigma}, \boldsymbol{\alpha}) = \frac{2 \exp\{(\mathbf{x} - \boldsymbol{\mu})'\boldsymbol{\Sigma}^{-1}\boldsymbol{\alpha}\}}{(2\pi)^{p/2}|\boldsymbol{\Sigma}|^{1/2}} \left(\frac{\delta(\mathbf{x}, \boldsymbol{\mu} \mid \boldsymbol{\Sigma})}{2 + \boldsymbol{\alpha}'\boldsymbol{\Sigma}^{-1}\boldsymbol{\alpha}} \right)^{\lambda/2} K_\lambda(u), \quad (6.7)$$

where $u = \sqrt{(2 + \boldsymbol{\alpha}'\boldsymbol{\Sigma}^{-1}\boldsymbol{\alpha})\delta(\mathbf{x}, \boldsymbol{\mu} \mid \boldsymbol{\Sigma})}$, $\delta(\mathbf{x}, \boldsymbol{\mu} \mid \boldsymbol{\Sigma}) = (\mathbf{x} - \boldsymbol{\mu})'\boldsymbol{\Sigma}^{-1}(\mathbf{x} - \boldsymbol{\mu})$, and λ, $\boldsymbol{\alpha}$, and $\boldsymbol{\Sigma}$ are defined as before.

Now the random variable $\mathbf{X} \sim \mathrm{SAL}_p(\boldsymbol{\mu}, \boldsymbol{\Sigma}, \boldsymbol{\alpha})$ can be generated via

$$\mathbf{X} = \boldsymbol{\mu} + W\boldsymbol{\alpha} + \sqrt{W}\mathbf{Y}, \qquad (6.8)$$

where $\mathbf{Y} \sim \mathrm{N}(\mathbf{0}, \boldsymbol{\Sigma})$, and so $\mathbf{X} \mid w \sim \mathrm{N}(\boldsymbol{\mu} + w\boldsymbol{\alpha}, w\boldsymbol{\Sigma})$. Bayes' theorem gives

$$f_W(w \mid \mathbf{x}) = \frac{f_{\mathbf{X}}(\mathbf{x} \mid w)h(w)}{f_{\mathbf{X}}(\mathbf{x})}, \qquad (6.9)$$

where $\mathbf{X} \mid w \sim \mathrm{N}(\boldsymbol{\mu} + w\boldsymbol{\alpha}, w\boldsymbol{\Sigma})$, $W \backsim \mathrm{Exp}(1)$, and $f_{\mathbf{X}}(\mathbf{x})$ is the density of the SAL distribution in (6.7). Note that $W \backsim \mathrm{Exp}(1)$ denotes that the random variable W follows an exponential distribution with rate 1. From (6.9),

$$\begin{aligned}
f_W(w \mid \mathbf{x}) = {} & \frac{w^{\lambda-1}}{2} \left(\frac{\delta(\mathbf{x}, \boldsymbol{\mu} \mid \boldsymbol{\Sigma})}{2 + \boldsymbol{\alpha}'\boldsymbol{\Sigma}^{-1}\boldsymbol{\alpha}} \right)^{-\lambda/2} \\
& \times \frac{\exp\{-\frac{1}{2w}\delta(\mathbf{x}, \boldsymbol{\mu} \mid \boldsymbol{\Sigma}) - \frac{w}{2}(2 + \boldsymbol{\alpha}'\boldsymbol{\Sigma}^{-1}\boldsymbol{\alpha})\}}{K_\lambda\left(\sqrt{(2 + \boldsymbol{\alpha}'\boldsymbol{\Sigma}^{-1}\boldsymbol{\alpha})\delta(\mathbf{x}, \boldsymbol{\mu} \mid \boldsymbol{\Sigma})}\right)}
\end{aligned} \qquad (6.10)$$

is the density of a GIG random variable with parameters $a = 2 + \boldsymbol{\alpha}'\boldsymbol{\Sigma}^{-1}\boldsymbol{\alpha}$

and $b = \delta(\mathbf{x}, \boldsymbol{\mu} \mid \boldsymbol{\Sigma})$, cf. (6.1). Franczak et al. (2014) develop a finite mixture of SAL distributions with density

$$f(\mathbf{x} \mid \boldsymbol{\vartheta}) = \sum_{g=1}^{G} \pi_g f_{\mathrm{SAL}}(\mathbf{x} \mid \boldsymbol{\mu}_g, \boldsymbol{\Sigma}_g, \boldsymbol{\alpha}_g),$$

where $\boldsymbol{\vartheta}$ denotes all model parameters.

6.3.2   Parameter Estimation

For SAL mixture models, the complete-data comprise the observed $\mathbf{x}_i$ together with the missing component membership labels $\mathbf{z}_i$ and the latent w_{ig}, for $i = 1, \ldots, n$ and $g = 1, \ldots, G$. Define $\mathbf{z}_i = (z_{i1}, \ldots, z_{iG})$ in the usual way so that $z_{ig} = 1$ if observation i belongs to component g and $z_{ig} = 0$ otherwise, for $i = 1, \ldots, n$ and $g = 1, \ldots, G$. Noting that

$$\mathbf{X}_i \mid w_{ig}, z_{ig} = 1 \sim \mathrm{N}(\boldsymbol{\mu}_g + w_{ig}\boldsymbol{\alpha}_g, w_{ig}\boldsymbol{\Sigma}_g)$$

independently, for $i = 1, \ldots, n$, and $W_{ig} \mid z_{ig} = 1 \sim \mathrm{Exp}(1)$, the complete-data likelihood is given by

$$\mathcal{L}_c = \prod_{i=1}^{n} \prod_{g=1}^{G} \left[\pi_g \phi \left(\mathbf{x}_i \mid \boldsymbol{\mu}_g + w_{ig}\boldsymbol{\alpha}_g, w_{ig}\boldsymbol{\Sigma}_g \right) h\left(w_{ig}\right) \right]^{z_{ig}},$$

where $h(w_{ig})$ is the density of an exponential distribution with rate 1. Note that $W_{ig} \mid \mathbf{x}_i, z_{ig} = 1 \sim \mathrm{GIG}(a_g, b_{ig}, \lambda)$, where $a_g = 2 + \boldsymbol{\alpha}_g' \boldsymbol{\Sigma}_g^{-1} \boldsymbol{\alpha}_g$ and $b_{ig} = \delta(\mathbf{x}_i, \boldsymbol{\mu}_g \mid \boldsymbol{\Sigma}_g)$. Then the expected value of the complete-data log-likelihood is

$$\mathcal{Q} = C + \sum_{g=1}^{G} n_g \log \pi_g + \sum_{g=1}^{G} \frac{n_g}{2} \log \left| \boldsymbol{\Sigma}_g^{-1} \right| - \sum_{i=1}^{n} \sum_{g=1}^{G} \hat{z}_{ig} E_{1ig}$$

$$- \frac{1}{2} \sum_{i=1}^{n} \sum_{g=1}^{G} \hat{z}_{ig} E_{2ig} \left(\mathbf{x}_i - \boldsymbol{\mu}_g \right)' \boldsymbol{\Sigma}_g^{-1} \left(\mathbf{x}_i - \boldsymbol{\mu}_g \right)$$

$$+ \sum_{i=1}^{n} \sum_{g=1}^{G} \hat{z}_{ig} \left(\mathbf{x}_i - \boldsymbol{\mu}_g \right)' \boldsymbol{\Sigma}_g^{-1} \boldsymbol{\alpha}_g - \frac{1}{2} \sum_{i=1}^{n} \sum_{g=1}^{G} \hat{z}_{ig} E_{1ig} \boldsymbol{\alpha}_g' \boldsymbol{\Sigma}_g^{-1} \boldsymbol{\alpha}_g,$$

where C is constant with respect to the model parameters, $n_g = \sum_{i=1}^{n} \hat{z}_{ig}$,

$$\hat{z}_{ig} = \mathbb{E}\left[Z_{ig} \mid \mathbf{x}_i\right] = \frac{\hat{\pi}_g f_{\mathrm{SAL}}(\mathbf{x}_i \mid \hat{\boldsymbol{\mu}}_g, \hat{\boldsymbol{\Sigma}}_g, \hat{\boldsymbol{\alpha}}_g)}{\sum_{h=1}^{G} \hat{\pi}_h f_{\mathrm{SAL}}(\mathbf{x}_i \mid \hat{\boldsymbol{\mu}}_h, \hat{\boldsymbol{\Sigma}}_h, \hat{\boldsymbol{\alpha}}_h)}, \tag{6.11}$$

and, because $W_{ig} \mid \mathbf{x}_i, z_{ig} = 1 \sim \mathrm{GIG}(a_g, b_{ig}, \lambda)$, we can use (6.2) and (6.3) to write

$$E_{1ig} := \mathbb{E}\left[W_{ig} \mid \mathbf{x}_i, z_{ig} = 1\right] = \sqrt{\frac{b_{ig}}{a_g}} \frac{K_{\lambda+1}\left(\sqrt{a_g b_{ig}}\right)}{K_{\lambda}\left(\sqrt{a_g b_{ig}}\right)}, \tag{6.12}$$

and

$$E_{2ig} := \mathbb{E}\left[1/W_{ig} \mid \mathbf{x}_i, z_{ig} = 1\right] = \sqrt{\frac{a_g}{b_{ig}}} \frac{K_{\lambda+1}\left(\sqrt{a_g b_{ig}}\right)}{K_\lambda\left(\sqrt{a_g b_{ig}}\right)} - \frac{2\lambda}{b_{ig}}. \quad (6.13)$$

Note that, as always, the expected values (6.11)–(6.13) are conditional on the current parameter estimates.

The expected value of the complete-data log-likelihood, i.e., $\mathcal{Q}$, is maximized to obtain the updates $\hat{\pi}_g = n_g/n$,

$$\hat{\boldsymbol{\alpha}}_g = \frac{\left(\sum_{i=1}^n \hat{z}_{ig} E_{2ig}\right)\left(\sum_{j=1}^n \hat{z}_{jg} \mathbf{x}_j\right) - n_g \sum_{i=1}^n \hat{z}_{ig} E_{2ig} \mathbf{x}_i}{\left(\sum_{i=1}^n \hat{z}_{ig} E_{1ig}\right)\left(\sum_{j=1}^n \hat{z}_{jg} E_{2jg}\right) - n_g^2}, \quad (6.14)$$

$$\hat{\boldsymbol{\mu}}_g = \frac{\left(\sum_{i=1}^n \hat{z}_{ig} E_{1ig}\right)\left(\sum_{j=1}^n \hat{z}_{jg} E_{2jg} \mathbf{x}_j\right) - n_g \sum_{i=1}^n \hat{z}_{ig} \mathbf{x}_i}{\left(\sum_{i=1}^n \hat{z}_{ig} E_{1ig}\right)\left(\sum_{j=1}^n \hat{z}_{jg} E_{2jg}\right) - n_g^2}, \quad (6.15)$$

and

$$\hat{\boldsymbol{\Sigma}}_g = \mathbf{S}_g - \hat{\boldsymbol{\alpha}}_g \mathbf{r}_g' - \mathbf{r}_g \hat{\boldsymbol{\alpha}}_g' + \frac{1}{n_g} \hat{\boldsymbol{\alpha}}_g \hat{\boldsymbol{\alpha}}_g' \sum_{i=1}^n \hat{z}_{ig} E_{1ig}, \quad (6.16)$$

where

$$\mathbf{S}_g = \frac{1}{n_g} \sum_{i=1}^n \hat{z}_{ig} E_{2ig} \left(\mathbf{x}_i - \hat{\boldsymbol{\mu}}_g\right) \left(\mathbf{x}_i - \hat{\boldsymbol{\mu}}_g\right)'$$

and

$$\mathbf{r}_g = \frac{1}{n_g} \sum_{i=1}^n \hat{z}_{ig} \left(\mathbf{x}_i - \hat{\boldsymbol{\mu}}_g\right).$$

Further details are available in Franczak et al. (2014). Now it is possible to summarize this EM algorithm for the mixture of SAL distributions.

EM Algorithm for SAL Mixture

initialize $\hat{\pi}_g$, $\hat{\boldsymbol{\alpha}}_g$, $\hat{\boldsymbol{\mu}}_g$, and $\hat{\boldsymbol{\Sigma}}_g$, e.g., via deterministic annealing
while convergence criterion not met
    update $\hat{z}_{ig}$ according to (6.11)
    update E_{1ig} according to (6.12)
    update E_{2ig} according to (6.13)
    update $\hat{\pi}_g = n_g/n$
    update $\hat{\boldsymbol{\alpha}}_g$ according to (6.14)
    update $\hat{\boldsymbol{\mu}}_g$ according to (6.15)
    update $\hat{\boldsymbol{\Sigma}}_g$ according to (6.16)
    check convergence criterion
end while

Note that Franczak et al. (2014) use deterministic annealing to initialize the algorithm. They also report that the ICL may be more effective than the BIC for selecting the number of components in a SAL mixture.

6.3.3   SAL Mixtures versus Gaussian Mixtures

Before proceeding to mixtures of more flexible asymmetric distributions, it is useful to consider the relative performance of SAL mixtures and Gaussian mixtures when clusters are asymmetric. First, consider one component from a SAL distribution (Figure 6.1). Fitting a Gaussian mixture to this component, within a clustering paradigm, a mechanism emerges by which a Gaussian mixture can be used to capture an asymmetric cluster via multiple components (Figure 6.1). Situations such as this are reminiscent of the flame on a candle: one Gaussian distribution is fitted to the part nearest the wick, another is fitted to the luminous zone, and a third is fitted to the veil — or roughly so. Whether this approach will work for multiple asymmetric clusters will depend, *inter alia*, on how well the clusters are separated.

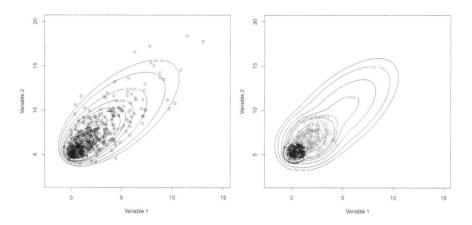

Figure 6.1
Scatter plots of data from a SAL distribution, with contours from a fitted SAL distribution (left) and contours from a fitted $G = 3$ component Gaussian mixture model (right), where plotting symbols and shading represent predicted classifications.

Consider the data in Figure 6.2, where there are two asymmetric clusters that can be separated by a straight line. These data are generated from a $G = 2$ component SAL mixture and so it is not surprising that fitting SAL mixtures to these data leads to the selection of a $G = 2$ component SAL mixture with perfect class agreement (Figure 6.2). Gaussian mixtures are fitted to these data for $G = 1, \ldots, 6$ components and the BIC selects a $G = 5$ component model; here the Gaussian components cannot be merged to return the correct clusters because one (high-variance) Gaussian component has been used to capture all points that do not better fit within one of the other four components (Figure 6.3). While this is obvious by inspection in two dimensions, it would be difficult to detect in higher dimensions. The unsuitability of Gaussian mixtures for capturing asymmetric clusters via *a posteriori* merging has been noted previously (e.g., Franczak et al., 2014; Murray et al., 2014a). This is one reason why it has been said that merging Gaussian components is not a "get out of jail free" card (McNicholas and Browne, 2013).

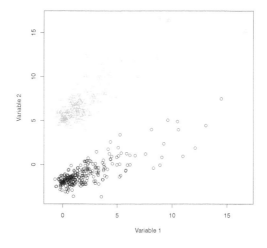

Figure 6.2
Scatter plot of a two-component mixture of SAL distributions, where plotting symbol and shading reflect component membership.

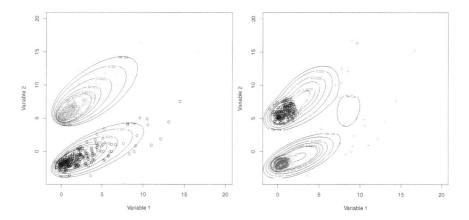

Figure 6.3
Scatter plots depicting the data from Figure 6.2, with contours from a fitted $G = 2$ component SAL mixture (left) and contours from a $G = 5$ component Gaussian mixture (right), where plotting symbols and shading represent predicted classifications in each case.

6.4 Mixture of Generalized Hyperbolic Distributions

6.4.1 Generalized Hyperbolic Distribution

From McNeil et al. (2005), the density of a random variable $\mathbf{X}$ following the generalized hyperbolic distribution can be written

$$f(\mathbf{x} \mid \boldsymbol{\theta}) = \left[\frac{\chi + \delta\left(\mathbf{x}, \boldsymbol{\mu} \mid \boldsymbol{\Delta}\right)}{\psi + \boldsymbol{\gamma}'\boldsymbol{\Delta}^{-1}\boldsymbol{\gamma}} \right]^{(\lambda-p/2)/2}$$
$$\times \frac{[\psi/\chi]^{\lambda/2} K_{\lambda-p/2}\left(\sqrt{[\psi + \boldsymbol{\gamma}'\boldsymbol{\Delta}^{-1}\boldsymbol{\gamma}][\chi + \delta(\mathbf{x}, \boldsymbol{\mu} \mid \boldsymbol{\Delta})]}\right)}{(2\pi)^{p/2} |\boldsymbol{\Delta}|^{1/2} K_\lambda\left(\sqrt{\chi\psi}\right) \exp\left\{(\boldsymbol{\mu} - \mathbf{x})'\boldsymbol{\Delta}^{-1}\boldsymbol{\gamma}\right\}}, \tag{6.17}$$

where $\boldsymbol{\mu}$ is the location parameter, $\boldsymbol{\gamma}$ is the skewness parameter, $\boldsymbol{\Delta}$ is the scale matrix, such that $|\boldsymbol{\Delta}| = 1$, λ is the index parameter, ψ and χ are concentration parameters, $\delta\left(\mathbf{x}, \boldsymbol{\mu} \mid \boldsymbol{\Delta}\right) = (\mathbf{x} - \boldsymbol{\mu})'\boldsymbol{\Delta}^{-1}(\mathbf{x} - \boldsymbol{\mu})$, and $\boldsymbol{\theta} = (\boldsymbol{\mu}, \boldsymbol{\Delta}, \boldsymbol{\gamma}, \lambda, \chi, \psi)$ is the vector of parameters. The notation $\mathbf{X} \sim G_p\left(\boldsymbol{\mu}, \boldsymbol{\Delta}, \boldsymbol{\gamma}, \chi, \psi, \lambda\right)$ is used to indicate that the p-dimensional random variable $\mathbf{X}$ has the generalized hyperbolic density in (6.17). As Browne and McNicholas (2015) point out, such a random variable $\mathbf{X}$ can be generated by combining a random variable $W \sim \text{GIG}(\psi, \chi, \lambda)$ and a latent multivariate Gaussian random variable $\mathbf{V} \sim N(\mathbf{0}, \boldsymbol{\Delta})$ via the relationship

$$\mathbf{X} = \boldsymbol{\mu} + W\boldsymbol{\gamma} + \sqrt{W}\mathbf{V}, \tag{6.18}$$

and it follows that $\mathbf{X} \mid w \sim N(\boldsymbol{\mu} + w\boldsymbol{\gamma}, w\boldsymbol{\Delta})$. Note that the relationship in (6.18) is analogous to that used for the SAL distribution in (6.8), and, similar to the SAL distribution, it follows from Bayes' theorem that

$$f(w \mid \mathbf{x}) = \frac{f(\mathbf{x} \mid w)h(w)}{f(\mathbf{x})} = \left[\frac{\psi + \boldsymbol{\gamma}'\boldsymbol{\Delta}^{-1}\boldsymbol{\gamma}}{\chi + \delta\left(\mathbf{x}, \boldsymbol{\mu} \mid \boldsymbol{\Delta}\right)} \right]^{(\lambda-p/2)/2}$$
$$\times \frac{w^{\lambda-p/2-1} \exp\left\{ -\left[w\left(\psi + \boldsymbol{\gamma}'\boldsymbol{\Delta}^{-1}\boldsymbol{\gamma}\right) + (\chi + \delta\left(\mathbf{x}, \boldsymbol{\mu} \mid \boldsymbol{\Delta}\right))/w\right]/2\right\}}{2K_{\lambda-p/2}\left(\sqrt{[\psi + \boldsymbol{\gamma}'\boldsymbol{\Delta}^{-1}\boldsymbol{\gamma}][\chi + \delta\left(\mathbf{x}, \boldsymbol{\mu} \mid \boldsymbol{\Delta}\right)]}\right)},$$

i.e.,

$$W \mid \mathbf{x} \sim \text{GIG}(\psi + \boldsymbol{\gamma}'\boldsymbol{\Delta}^{-1}\boldsymbol{\gamma}, \chi + \delta\left(\mathbf{x}, \boldsymbol{\mu} \mid \boldsymbol{\Delta}\right), \lambda - p/2).$$

The generalized hyperbolic distribution has a number of well-known special and limiting cases (cf. McNeil et al., 2005), and some of these are illustrated in Figure 6.4. Note that Figure 6.4 is intended as a rough sketch and, in practice, further considerations are needed for some of the models depicted therein; details on the generalized hyperbolic distribution are given here and details about the normal-inverse Gaussian distribution are provided by Karlis and Santourian (2009).

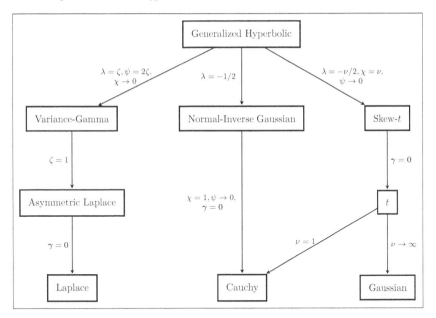

Figure 6.4
Several distributions available as special and limiting cases of the generalized hyperbolic distribution.

With classification and clustering applications in mind, it is desirable to relax the condition $|\mathbf{\Delta}| = 1$. As Hu (2005) notes, relaxing this condition results in an identifiability issue because the densities of $\mathbf{X}_1 \sim G_p(\mathbf{\mu}, c\mathbf{\Sigma}, c\mathbf{\gamma}, \chi/c, c\psi, \lambda)$ and $\mathbf{X}_2 \sim G_p(\mathbf{\mu}, \mathbf{\Sigma}, \mathbf{\gamma}, \chi, \psi, \lambda)$ are identical for any $c \in \mathbb{R}^+$, where $\mathbf{\Sigma}$ denotes the unconstrained scale matrix. Browne and McNicholas (2015) overcome this issue by writing $\omega = \sqrt{\psi\chi}$ and $\eta = \sqrt{\chi/\psi}$, and using the relationship in (6.18) with $\eta = 1$. That is,

$$\mathbf{X} = \mathbf{\mu} + W\eta\mathbf{\gamma} + \sqrt{W\eta}\mathbf{V} = \mathbf{\mu} + W\mathbf{\alpha} + \sqrt{W}\mathbf{V}, \tag{6.19}$$

where $\mathbf{\alpha} = \eta\mathbf{\gamma} = \mathbf{\gamma}$, $\mathbf{V} \sim N(\mathbf{0}, \mathbf{\Sigma})$, and $W \sim I(\omega, 1, \lambda)$, cf. (6.5). Under this parameterization, the density of the generalized hyperbolic distribution is

$$f_H(\mathbf{x} \mid \mathbf{\theta}) = \left[\frac{\omega + \delta(\mathbf{x}, \mathbf{\mu}|\mathbf{\Sigma})}{\omega + \mathbf{\alpha}'\mathbf{\Sigma}^{-1}\mathbf{\alpha}}\right]^{(\lambda-p/2)/2}$$
$$\times \frac{K_{\lambda-p/2}\left(\sqrt{[\omega + \mathbf{\alpha}'\mathbf{\Sigma}^{-1}\mathbf{\alpha}][\omega + \delta(\mathbf{x}, \mathbf{\mu}|\mathbf{\Sigma})]}\right)}{(2\pi)^{p/2}|\mathbf{\Sigma}|^{1/2}K_\lambda(\omega)\exp\{-(\mathbf{x} - \mathbf{\mu})'\mathbf{\Sigma}^{-1}\mathbf{\alpha}\}}, \tag{6.20}$$

where $\mathbf{\theta} = (\mathbf{\mu}, \mathbf{\Sigma}, \mathbf{\alpha}, \omega, \lambda)$, $\mathbf{\mu}$ is the location parameter, $\mathbf{\alpha}$ is the skewness parameter, $\mathbf{\Sigma}$ is the scale matrix, λ is the index parameter, and ω is the concentration parameter. Now $\mathbf{X} \mid w \sim N(\boldsymbol{\mu} + w\boldsymbol{\alpha}, w\boldsymbol{\Sigma})$ and

$$W \mid \mathbf{x} \sim \text{GIG}(\omega + \boldsymbol{\alpha}'\boldsymbol{\Sigma}^{-1}\boldsymbol{\alpha}, \omega + \delta\left(\mathbf{x}, \boldsymbol{\mu}|\boldsymbol{\Sigma}\right), \lambda - p/2).$$

Write $\mathbf{X} \sim \text{GHD}(\boldsymbol{\mu}, \boldsymbol{\Sigma}, \boldsymbol{\alpha}, \omega, \lambda)$ to denote that the p-dimensional random variable $\mathbf{X}$ has the density in (6.20).

Browne and McNicholas (2015) use a finite mixture of generalized hyperbolic distributions (MGHD) with density

$$f(\mathbf{x} \mid \boldsymbol{\vartheta}) = \sum_{g=1}^{G} \pi_g f_{\text{H}}(\mathbf{x} \mid \boldsymbol{\theta}_g), \tag{6.21}$$

where $\boldsymbol{\vartheta}$ is the vector of all model parameters and $f_{\text{H}}(\mathbf{x} \mid \boldsymbol{\theta}_g)$ is the density in (6.20) with $\boldsymbol{\theta}_g = (\boldsymbol{\mu}_g, \boldsymbol{\Sigma}_g, \boldsymbol{\alpha}_g, \omega_g, \lambda_g)$.

6.4.2   Parameter Estimation

Suppose that $\mathbf{x}_1, \ldots, \mathbf{x}_n$ are observed from a generalized hyperbolic mixture (6.21) within a clustering paradigm so that all observations are unlabelled. The complete-data comprise the observed $\mathbf{x}_i$ together with the missing component membership labels $\mathbf{z}_i$ and the latent w_{ig}, for $i = 1, \ldots, n$ and $g = 1, \ldots, G$. Define $\mathbf{z}_i = (z_{i1}, \ldots, z_{iG})$ in the usual way so that $z_{ig} = 1$ if observation i belongs to component g and $z_{ig} = 0$ otherwise.

Noting that $\mathbf{X}_i \mid w_{ig}, z_{ig} = 1 \sim N(\boldsymbol{\mu}_g + w_{ig}\boldsymbol{\alpha}_g, w_{ig}\boldsymbol{\Sigma}_g)$ independently for $i = 1, \ldots, n$ and $W_{ig} \mid z_{ig} = 1 \sim \text{I}(\omega_g, 1, \lambda_g)$, the complete-data log-likelihood is given by

$$l_{\text{c}}(\boldsymbol{\vartheta}) = C - \frac{1}{2} \sum_{i=1}^{n} \sum_{g=1}^{G} z_{ig} \log \left| \boldsymbol{\Sigma}_g^{-1} \right| + \sum_{i=1}^{n} \sum_{g=1}^{G} z_{ig} \log h(w_{ig} \mid \omega_g, 1, \lambda_g)$$

$$- \frac{1}{2} \text{tr} \left\{ \sum_{g=1}^{G} \boldsymbol{\Sigma}_g^{-1} \sum_{i=1}^{n} z_{ig} \left[\frac{1}{w_{ig}} (\mathbf{x}_i - \boldsymbol{\mu}_g)(\mathbf{x}_i - \boldsymbol{\mu}_g)' \right. \right.$$

$$\left. \left. - (\mathbf{x}_i - \boldsymbol{\mu}_g)\boldsymbol{\alpha}_g' - \boldsymbol{\alpha}_g(\mathbf{x}_i - \boldsymbol{\mu}_g)' + w_{ig}\boldsymbol{\alpha}_g\boldsymbol{\alpha}_g' \right] \right\},$$

where C does not depend on the model parameters and $h(w_{ig} \mid \omega_g, 1, \lambda_g)$ is the GIG density defined in (6.5). Because

$$W_{ig} \mid \mathbf{x}_i, z_{ig} = 1 \sim \text{GIG}(\omega_g + \boldsymbol{\alpha}_g'\boldsymbol{\Sigma}_g^{-1}\boldsymbol{\alpha}_g, \omega_g + \delta(\mathbf{x}_i, \boldsymbol{\mu}_g|\boldsymbol{\Sigma}_g), \lambda_g - p/2),$$

the expectations given in (6.2)–(6.4) can be used in the E-step. Specifically, in addition to the familiar expectation

$$\hat{z}_{ig} := \mathbb{E}\left[Z_{ig} \mid \mathbf{x}_i\right] = \frac{\hat{\pi}_g f_{\text{H}}(\mathbf{x}_i \mid \hat{\boldsymbol{\theta}}_g)}{\sum_{h=1}^{G} \hat{\pi}_h f_{\text{H}}(\mathbf{x}_i \mid \hat{\boldsymbol{\theta}}_h)}, \tag{6.22}$$

we have

$$a_{ig} := \mathbb{E}\left[W_{ig} \mid \mathbf{x}_i, z_{ig} = 1\right] = \sqrt{\frac{\hat{\omega}_g + \delta(\mathbf{x}_i, \hat{\boldsymbol{\mu}}_g | \hat{\boldsymbol{\Sigma}}_g)}{\hat{\omega}_g + \hat{\boldsymbol{\alpha}}_g' \hat{\boldsymbol{\Sigma}}_g^{-1} \hat{\boldsymbol{\alpha}}_g}}$$

$$\times \frac{K_{\hat{\lambda}_g - p/2 + 1}\left(\sqrt{(\hat{\omega}_g + \hat{\boldsymbol{\alpha}}_g' \hat{\boldsymbol{\Sigma}}_g^{-1} \hat{\boldsymbol{\alpha}}_g)(\hat{\omega}_g + \delta(\mathbf{x}_i, \hat{\boldsymbol{\mu}}_g | \hat{\boldsymbol{\Sigma}}_g))}\right)}{K_{\hat{\lambda}_g - p/2}\left(\sqrt{(\hat{\omega}_g + \hat{\boldsymbol{\alpha}}_g' \hat{\boldsymbol{\Sigma}}_g^{-1} \hat{\boldsymbol{\alpha}}_g)(\hat{\omega}_g + \delta(\mathbf{x}_i, \hat{\boldsymbol{\mu}}_g | \hat{\boldsymbol{\Sigma}}_g))}\right)}, \qquad (6.23)$$

$$b_{ig} := \mathbb{E}[1/W_{ig} \mid \mathbf{x}_i, z_{ig} = 1] = \frac{p - 2\hat{\lambda}_g}{\hat{\omega}_g + \delta(\mathbf{x}_i, \hat{\boldsymbol{\mu}}_g | \hat{\boldsymbol{\Sigma}}_g)} + \sqrt{\frac{\hat{\omega}_g + \hat{\boldsymbol{\alpha}}_g' \hat{\boldsymbol{\Sigma}}_g^{-1} \hat{\boldsymbol{\alpha}}_g}{\hat{\omega}_g + \delta(\mathbf{x}_i, \hat{\boldsymbol{\mu}}_g | \hat{\boldsymbol{\Sigma}}_g)}}$$

$$\times \frac{K_{\hat{\lambda}_g - p/2 + 1}\left(\sqrt{(\hat{\omega}_g + \hat{\boldsymbol{\alpha}}_g' \hat{\boldsymbol{\Sigma}}_g^{-1} \hat{\boldsymbol{\alpha}}_g)(\hat{\omega}_g + \delta(\mathbf{x}_i, \hat{\boldsymbol{\mu}}_g | \hat{\boldsymbol{\Sigma}}_g))}\right)}{K_{\hat{\lambda}_g - p/2}\left(\sqrt{(\hat{\omega}_g + \hat{\boldsymbol{\alpha}}_g' \hat{\boldsymbol{\Sigma}}_g^{-1} \hat{\boldsymbol{\alpha}}_g)(\hat{\omega}_g + \delta(\mathbf{x}_i, \hat{\boldsymbol{\mu}}_g | \hat{\boldsymbol{\Sigma}}_g))}\right)},$$

$$(6.24)$$

and

$$c_{ig} := \mathbb{E}\left[\log W_{ig} \mid \mathbf{x}_i, z_{ig} = 1\right] = \log \sqrt{\frac{\hat{\omega}_g + \delta(\mathbf{x}_i, \hat{\boldsymbol{\mu}}_g | \hat{\boldsymbol{\Sigma}}_g)}{\hat{\omega}_g + \hat{\boldsymbol{\alpha}}_g' \hat{\boldsymbol{\Sigma}}_g^{-1} \hat{\boldsymbol{\alpha}}_g}}$$

$$+ \frac{\partial}{\partial t} \log \left\{ K_t \left(\sqrt{(\hat{\omega}_g + \hat{\boldsymbol{\alpha}}_g' \hat{\boldsymbol{\Sigma}}_g^{-1} \hat{\boldsymbol{\alpha}}_g)(\hat{\omega}_g + \delta(\mathbf{x}_i, \hat{\boldsymbol{\mu}}_g | \hat{\boldsymbol{\Sigma}}_g))} \right) \right\} \Bigg|_{t = \hat{\lambda}_g - p/2}.$$

$$(6.25)$$

Note that, as always, the expected values (6.22)–(6.25) are computed based on the current parameter estimates. Following Browne and McNicholas (2015), the following notation is used:

$$\bar{a}_g = \frac{1}{n_g} \sum_{i=1}^n \hat{z}_{ig} a_{ig}, \quad \bar{b}_g = \frac{1}{n_g} \sum_{i=1}^n \hat{z}_{ig} b_{ig}, \quad \text{and} \quad \bar{c}_g = \frac{1}{n_g} \sum_{i=1}^n \hat{z}_{ig} c_{ig},$$

where $n_g = \sum_{i=1}^n \hat{z}_{ig}$.

In the M-step, Browne and McNicholas (2015) use the following updates: $\hat{\pi}_g = n_g/n$,

$$\hat{\boldsymbol{\mu}}_g = \frac{\sum_{i=1}^n \hat{z}_{ig} \mathbf{x}_i (\bar{a}_g b_{ig} - 1)}{\sum_{i=1}^n \hat{z}_{ig} (\bar{a}_g b_{ig} - 1)},$$

$$\hat{\boldsymbol{\alpha}}_g = \frac{\sum_{i=1}^n \hat{z}_{ig} \mathbf{x}_i (\bar{b}_g - b_{ig})}{\sum_{i=1}^n \hat{z}_{ig} (\bar{a}_g b_{ig} - 1)},$$

$$\hat{\boldsymbol{\Sigma}}_g = \frac{1}{n_g} \sum_{i=1}^n \hat{z}_{ig} b_{ig} (\mathbf{x}_i - \hat{\boldsymbol{\mu}}_g)(\mathbf{x}_i - \hat{\boldsymbol{\mu}}_g)' - \hat{\boldsymbol{\alpha}}_g (\bar{\mathbf{x}}_g - \hat{\boldsymbol{\mu}}_g)'$$

$$- (\bar{\mathbf{x}}_g - \hat{\boldsymbol{\mu}}_g)(\hat{\boldsymbol{\alpha}}_g)' + \bar{a}_g \hat{\boldsymbol{\alpha}}_g (\hat{\boldsymbol{\alpha}}_g)',$$

$$\hat{\lambda}_g^{\mathrm{new}} = \bar{c}_g \hat{\lambda}_g \left[\frac{\partial}{\partial t} \log K_t \left(\hat{\omega}_g \right) \Big|_{t = \hat{\lambda}_g} \right]^{-1},$$

$$\hat{\omega}_g^{\mathrm{new}} = \hat{\omega}_g - \left[\frac{\partial}{\partial t} q_g \left(t, \hat{\lambda}_g^{\mathrm{new}} \right) \Big|_{t = \hat{\omega}_g} \right] \left[\frac{\partial^2}{\partial t^2} q_g \left(t, \hat{\lambda}_g^{\mathrm{new}} \right) \Big|_{t = \hat{\omega}_g} \right]^{-1},$$

where $\bar{\mathbf{x}}_g = (1/n_g) \sum_{i=1}^n \hat{z}_{ig} \mathbf{x}_i$ and

$$q_g(\omega_g, \lambda_g) = -\log K_{\lambda_g} (\omega_g) + (\lambda_g - 1)\bar{c}_g - \frac{\omega_g}{2} \left(\bar{a}_g + \bar{b}_g \right). \tag{6.26}$$

Browne and McNicholas (2015) show that $\hat{\boldsymbol{\Sigma}}_g$ is positive-definite. Note that the updates $\hat{\omega}_g^{\mathrm{new}}$ and $\hat{\lambda}_g^{\mathrm{new}}$ arise from maximizing the function (6.26) via conditional maximization, using the fact that $K_\lambda(\omega)$ is strictly log-convex with respect to λ and ω (Baricz, 2010); see Browne and McNicholas (2015) for details.

Now it is possible to summarize the EM algorithm for the MGHD.

EM Algorithm for MGHD

initialize $\hat{\pi}_g$, $\hat{\boldsymbol{\mu}}_g$, $\hat{\boldsymbol{\alpha}}_g$, $\hat{\boldsymbol{\Sigma}}_g$, $\hat{\lambda}_g$, $\hat{\omega}_g$
while convergence criterion not met
    update $\hat{z}_{ig}$
    update a_{ig}, b_{ig}, c_{ig}
    update $\hat{\pi}_g$, $\hat{\boldsymbol{\mu}}_g$, and $\hat{\boldsymbol{\alpha}}_g$
    update $\hat{\boldsymbol{\Sigma}}_g$
    update $\hat{\lambda}_g^{\mathrm{new}}$
    update $\hat{\omega}_g^{\mathrm{new}}$
    check convergence criterion
    $\hat{\lambda}_g \leftarrow \hat{\lambda}_g^{\mathrm{new}}$, $\hat{\omega}_g \leftarrow \hat{\omega}_g^{\mathrm{new}}$
end while

Note that, from (6.19), the mean and variance of $\mathbf{X}_i \mid z_{ig} = 1$ from the MGHD (6.21) are given by

$$\mathbb{E}[\mathbf{X}_i \mid z_{ig} = 1] = \boldsymbol{\mu}_g + \mathbb{E}[W_{ig} \mid z_{ig} = 1]\boldsymbol{\alpha}_g, \tag{6.27}$$

and

$$\mathbb{V}\mathrm{ar}[\mathbf{X}_i \mid z_{ig} = 1] = \mathbb{E}[W_{ig} \mid z_{ig} = 1]\boldsymbol{\Sigma}_g + \mathbb{V}\mathrm{ar}[W_{ig} \mid z_{ig} = 1]\boldsymbol{\alpha}_g \boldsymbol{\alpha}_g', \tag{6.28}$$

respectively. Now $\mathbb{E}[\mathbf{X}_i \mid z_{ig} = 1]$ can be interpreted as the component mean and $\mathbb{V}\mathrm{ar}[\mathbf{X}_i \mid z_{ig} = 1]$ as the component variance, or component covariance matrix. Neither quantity is needed for model-based clustering or classification using the MGHD; however, both are needed for the approach of Morris and McNicholas (2016), which combines dimension reduction with clustering or classification using the MGHD.

6.5 Mixture of Generalized Hyperbolic Factor Analyzers

6.5.1 The Model

Tortora et al. (2015b) extend the mixture of factor analyzers model to the generalized hyperbolic distribution by following the approach that Murray et al. (2014a) took for the mixture of skew-t factor analyzers. First, note that $\mathbf{V}$ in (6.19) can be decomposed using a factor analysis model, i.e.,

$$\mathbf{V} = \mathbf{\Lambda U} + \varepsilon,$$

where $\mathbf{U} \sim \mathrm{N}(\mathbf{0}, \mathbf{I}_q)$ and $\varepsilon \sim \mathrm{N}(\mathbf{0}, \mathbf{\Psi})$ in the usual fashion, cf. Section 3.1. The resulting model is

$$\mathbf{X} = \boldsymbol{\mu} + W\boldsymbol{\alpha} + \sqrt{W}(\mathbf{\Lambda U} + \varepsilon), \tag{6.29}$$

where $W \sim \mathrm{I}(\omega, 1, \lambda)$, and so

$$\mathbf{X} \mid w \sim \mathrm{N}(\boldsymbol{\mu} + w\boldsymbol{\alpha}, w(\mathbf{\Lambda\Lambda'} + \mathbf{\Psi})).$$

Following this approach, Tortora et al. (2015b) arrive at a mixture of generalized hyperbolic factor analyzers (MGHFA) model with density

$$f(\mathbf{x} \mid \boldsymbol{\vartheta}) = \sum_{g=1}^{G} \pi_g f_{\mathrm{H}}(\mathbf{x} \mid \boldsymbol{\theta}_g), \tag{6.30}$$

where $f_{\mathrm{H}}(\mathbf{x} \mid \boldsymbol{\theta}_g)$ is as defined in (6.20) with $\boldsymbol{\theta}_g = (\boldsymbol{\mu}_g, \mathbf{\Lambda}_g\mathbf{\Lambda}_g' + \mathbf{\Psi}_g, \boldsymbol{\alpha}_g, \omega_g, \lambda_g)$ so that the scale matrix for component g is

$$\mathbf{\Sigma}_g = \mathbf{\Lambda}_g\mathbf{\Lambda}_g' + \mathbf{\Psi}_g.$$

6.5.2 Parameter Estimation

The AECM algorithm allows specification of different complete-data at each stage of the algorithm, and is used for parameter estimation. Suppose that $\mathbf{x}_1, \ldots, \mathbf{x}_n$ are observed from a MGHFA model (6.30) within a clustering paradigm so that all observations are unlabelled. The complete-data comprise the observed $\mathbf{x}_i$ together with the missing labels z_{ig}, the latent w_{ig}, and the latent factors $\mathbf{u}_{ig}$, for $i = 1, \ldots, n$ and $g = 1, \ldots, G$. The expected values denoted $\hat{z}_{ig}$, a_{ig}, b_{ig}, and c_{ig} in Section 6.4.2 are again needed here, and the quantities $\bar{a}_g$, $\bar{b}_g$, and $\bar{c}_g$ have the same meaning.

At the second stage of the algorithm, the latent factors $\mathbf{U}_{ig}$ are part of the complete-data and the following expectations are also needed:

$$\mathbb{E}[\mathbf{U}_{ig} \mid \mathbf{x}_i, z_{ig} = 1] = \hat{\boldsymbol{\beta}}_g(\mathbf{x}_i - \hat{\boldsymbol{\mu}}_g - a_{ig}\hat{\boldsymbol{\alpha}}_g) =: \mathbf{E}_{1ig},$$

$$\mathbb{E}[(1/W_{ig})\mathbf{U}_{ig} \mid \mathbf{x}_i, z_{ig} = 1] = \hat{\boldsymbol{\beta}}_g[b_{ig}(\mathbf{x}_i - \hat{\boldsymbol{\mu}}_g) - \hat{\boldsymbol{\alpha}}_g] =: \mathbf{E}_{2ig},$$

$$\mathbb{E}[(1/W_{ig})\mathbf{U}_{ig}\mathbf{U}'_{ig} \mid \mathbf{x}_i, z_{ig} = 1] = b_{ig}[\mathbf{I}_q - \hat{\boldsymbol{\beta}}_g\hat{\boldsymbol{\Lambda}}_g + \hat{\boldsymbol{\beta}}_g(\mathbf{x}_i - \hat{\boldsymbol{\mu}}_g)(\mathbf{x}_i - \hat{\boldsymbol{\mu}}_g)'\hat{\boldsymbol{\beta}}'_g]$$
$$- \hat{\boldsymbol{\beta}}_g[(\mathbf{x}_i - \hat{\boldsymbol{\mu}}_g)\hat{\boldsymbol{\alpha}}'_g + \hat{\boldsymbol{\alpha}}_g(\mathbf{x}_i - \hat{\boldsymbol{\mu}}_g)']\hat{\boldsymbol{\beta}}'_g + a_{ig}\hat{\boldsymbol{\beta}}_g\hat{\boldsymbol{\alpha}}_g\hat{\boldsymbol{\alpha}}'_g\hat{\boldsymbol{\beta}}'_g =: \mathbf{E}_{3ig},$$

where $\hat{\boldsymbol{\beta}}_g = \hat{\boldsymbol{\Lambda}}'_g(\hat{\boldsymbol{\Lambda}}_g\hat{\boldsymbol{\Lambda}}'_g + \hat{\boldsymbol{\Psi}}_g)^{-1}$. As always, expected values used in E-steps are conditional on current parameter estimates.

At the first stage of the AECM algorithm, the complete-data comprise the $\mathbf{x}_i$, z_{ig}, and w_{ig}, for $i = 1, \ldots, n$ and $g = 1, \ldots, G$, and the mixing proportions π_g, the component location parameters $\boldsymbol{\mu}_g$, the skewness parameters $\boldsymbol{\alpha}_g$, the concentration parameters ω_g, and the index parameters λ_g are updated. The complete-data log-likelihood at the first stage is

$$l_1 = \sum_{i=1}^n \sum_{g=1}^G z_{ig}\Big[\log \pi_g + \log \phi(\mathbf{x}_i \mid \boldsymbol{\mu}_g + w_{ig}\boldsymbol{\alpha}_g, w_{ig}(\boldsymbol{\Lambda}_g\boldsymbol{\Lambda}'_g + \boldsymbol{\Psi}_g)) \tag{6.31}$$
$$+ \log h(w_{ig} \mid \omega_g, 1, \lambda_g)\Big].$$

Tortora et al. (2015b) compute the (conditional) expected value of (6.5.2) and then compute the following updates: $\hat{\pi}_g = n_g/n$,

$$\hat{\boldsymbol{\mu}}_g = \frac{\sum_{i=1}^n \hat{z}_{ig}\mathbf{x}_i(\bar{a}_g b_{ig} - 1)}{\sum_{i=1}^n \hat{z}_{ig}(\bar{a}_g b_{ig} - 1)}, \tag{6.32}$$

$$\hat{\boldsymbol{\alpha}}_g = \frac{\sum_{i=1}^n \hat{z}_{ig}\mathbf{x}_i(b_{ig} - \bar{b}_g)}{\sum_{i=1}^n \hat{z}_{ig}(\bar{a}_g b_{ig} - 1)}, \tag{6.33}$$

$$\hat{\lambda}_g^{\text{new}} = \bar{c}_g\hat{\lambda}_g\left[\frac{\partial}{\partial t}\log K_t(\hat{\omega}_g)\Big|_{t=\hat{\lambda}_g}\right]^{-1}, \tag{6.34}$$

$$\hat{\omega}_g^{\text{new}} = \hat{\omega}_g - \left[\frac{\partial}{\partial t}q_g(t, \hat{\lambda}_g^{\text{new}})\Big|_{t=\hat{\omega}_g}\right]\left[\frac{\partial^2}{\partial t^2}q_g(t, \hat{\lambda}_g^{\text{new}})\Big|_{t=\hat{\omega}_g}\right]^{-1}, \tag{6.35}$$

where

$$q_g(\omega_g, \lambda_g) = -\log K_\lambda(\omega_g) + (\lambda_g - 1)\bar{c}_g - \frac{\omega_g}{2}(\bar{a}_g + \bar{b}_g).$$

At the second stage of the AECM algorithm, the complete-data comprise the $\mathbf{x}_i$, z_{ig}, w_{ig}, and $\mathbf{u}_{ig}$, for $i = 1, \ldots, n$ and $g = 1, \ldots, G$. At this stage, $\boldsymbol{\Lambda}_g$ and $\boldsymbol{\Psi}_g$ are updated and the complete-data log-likelihood can be written

$$l_2 = \sum_{i=1}^n \sum_{g=1}^G z_{ig}\Big[\log \pi_g + \log \phi(\mathbf{x}_i \mid \boldsymbol{\mu}_g + w_{ig}\boldsymbol{\alpha}_g + \boldsymbol{\Lambda}_g\mathbf{u}_{ig}, w_{ig}\boldsymbol{\Psi}_g)$$

$$+ \log \phi(\mathbf{u}_{ig} \mid \mathbf{0}, w_{ig}\mathbf{I}_q) + \log h(w_{ig} \mid \omega_g, 1, \lambda_g)\Big],$$

i.e.,

$$
\begin{aligned}
l_2 = C - \frac{1}{2} \sum_{i=1}^{n} \sum_{g=1}^{G} z_{ig} \Bigg[& \log|\boldsymbol{\Psi}_g| + \frac{1}{w_{ig}} \operatorname{tr}\{(\mathbf{x}_i - \boldsymbol{\mu}_g)(\mathbf{x}_i - \boldsymbol{\mu}_g)'\boldsymbol{\Psi}_g^{-1}\} \\
& - 2\operatorname{tr}\{(\mathbf{x}_i - \boldsymbol{\mu}_g)\boldsymbol{\alpha}_g'\boldsymbol{\Psi}_g^{-1}\} + w_{ig}\operatorname{tr}\{\boldsymbol{\alpha}_g\boldsymbol{\alpha}_g'\boldsymbol{\Psi}_g^{-1}\} + 2\operatorname{tr}\{\boldsymbol{\alpha}_g'\boldsymbol{\Psi}_g^{-1}\boldsymbol{\Lambda}_g\mathbf{u}_{ig}\} \\
& - 2\operatorname{tr}\left\{(\mathbf{x}_i - \boldsymbol{\mu}_g)'\boldsymbol{\Psi}_g^{-1}\boldsymbol{\Lambda}_g\left(\frac{1}{w_{ig}}\mathbf{u}_{ig}\right)\right\} + \operatorname{tr}\left\{\boldsymbol{\Lambda}_g\left(\frac{1}{w_{ig}}\mathbf{u}_{ig}\mathbf{u}_{ig}'\right)\boldsymbol{\Lambda}_g'\boldsymbol{\Psi}_g^{-1}\right\}\Bigg],
\end{aligned}
$$

where C is constant with respect to $\boldsymbol{\Lambda}_g$ and $\boldsymbol{\Psi}_g$. The (conditional) expected value of complete-data log-likelihood l_2 is

$$
\begin{aligned}
Q_2 = C - \frac{1}{2} \sum_{i=1}^{n} \sum_{g=1}^{G} \hat{z}_{ig} \Bigg[& \log|\boldsymbol{\Psi}_g| + b_{ig}\operatorname{tr}\{(\mathbf{x}_i - \hat{\boldsymbol{\mu}}_g)(\mathbf{x}_i - \hat{\boldsymbol{\mu}}_g)'\boldsymbol{\Psi}_g^{-1}\} \\
& - 2\operatorname{tr}\{(\mathbf{x}_i - \hat{\boldsymbol{\mu}}_g)\hat{\boldsymbol{\alpha}}_g'\boldsymbol{\Psi}_g^{-1}\} + a_{ig}\operatorname{tr}\{\hat{\boldsymbol{\alpha}}_g\hat{\boldsymbol{\alpha}}_g'\boldsymbol{\Psi}_g^{-1}\} + 2\operatorname{tr}\{\hat{\boldsymbol{\alpha}}_g'\boldsymbol{\Psi}_g^{-1}\boldsymbol{\Lambda}_g\mathbf{E}_{1ig}\} \\
& - 2\operatorname{tr}\{(\mathbf{x}_i - \hat{\boldsymbol{\mu}}_g)'\boldsymbol{\Psi}_g^{-1}\boldsymbol{\Lambda}_g\mathbf{E}_{2ig}\} + \operatorname{tr}\{\boldsymbol{\Lambda}_g\mathbf{E}_{3ig}\boldsymbol{\Lambda}_g'\boldsymbol{\Psi}_g^{-1}\}\Bigg],
\end{aligned}
$$

where C is constant with respect to $\boldsymbol{\Lambda}_g$ and $\boldsymbol{\Psi}_g$. Differentiating Q_2 with respect to $\boldsymbol{\Lambda}_g$ gives

$$
\begin{aligned}
S_1(\boldsymbol{\Lambda}_g, \boldsymbol{\Psi}_g) &= \frac{\partial Q_2}{\partial \boldsymbol{\Lambda}_g} \\
&= -\frac{1}{2} \sum_{i=1}^{n} \hat{z}_{ig} \left[2\boldsymbol{\Psi}_g^{-1}\hat{\boldsymbol{\alpha}}_g\mathbf{E}_{1ig}' - 2\boldsymbol{\Psi}_g^{-1}(\mathbf{x}_i - \hat{\boldsymbol{\mu}}_g)\mathbf{E}_{2ig}' + \boldsymbol{\Psi}_g^{-1}\boldsymbol{\Lambda}_g(\mathbf{E}_{3ig}' + \mathbf{E}_{3ig}) \right].
\end{aligned}
$$

Noting that $\mathbf{E}_{3ig}$ is a symmetric matrix, solving $S_1(\hat{\boldsymbol{\Lambda}}_g^{\text{new}}, \hat{\boldsymbol{\Psi}}_g^{\text{new}}) = \mathbf{0}$ gives

$$
\hat{\boldsymbol{\Lambda}}_g^{\text{new}} = \left\{ \sum_{i=1}^{n} \hat{z}_{ig} \left[(\mathbf{x}_i - \hat{\boldsymbol{\mu}}_g)\mathbf{E}_{2ig}' - \hat{\boldsymbol{\alpha}}_g\mathbf{E}_{1ig}' \right] \right\} \left\{ \sum_{i=1}^{n} \hat{z}_{ig}\mathbf{E}_{3ig} \right\}^{-1}. \tag{6.36}
$$

Differentiating Q_2 with respect to $\boldsymbol{\Psi}_g^{-1}$ gives

$$
\begin{aligned}
S_2(\boldsymbol{\Lambda}_g, \boldsymbol{\Psi}_g) &= \frac{\partial Q_2}{\partial \boldsymbol{\Psi}_g^{-1}} \\
&= \frac{n_g}{2}\boldsymbol{\Psi}_g - \frac{1}{2} \sum_{i=1}^{n} \hat{z}_{ig} \Big[b_{ig}(\mathbf{x}_i - \hat{\boldsymbol{\mu}}_g)(\mathbf{x}_i - \hat{\boldsymbol{\mu}}_g)' - 2\hat{\boldsymbol{\alpha}}_g(\mathbf{x}_i - \hat{\boldsymbol{\mu}}_g)' + a_{ig}\hat{\boldsymbol{\alpha}}_g\hat{\boldsymbol{\alpha}}_g' \\
&\qquad\qquad + 2\hat{\boldsymbol{\alpha}}_g\mathbf{E}_{1ig}'\boldsymbol{\Lambda}_g' - 2(\mathbf{x}_i - \hat{\boldsymbol{\mu}}_g)\mathbf{E}_{2ig}'\boldsymbol{\Lambda}_g' + \boldsymbol{\Lambda}_g\mathbf{E}_{3ig}\boldsymbol{\Lambda}_g' \Big],
\end{aligned}
$$

and solving $\operatorname{diag}\{S_2(\hat{\boldsymbol{\Lambda}}_g^{\text{new}}, \hat{\boldsymbol{\Psi}}_g^{\text{new}})\} = \mathbf{0}$ gives the update

$$
\begin{aligned}
\hat{\boldsymbol{\Psi}}_g^{\text{new}} = \frac{1}{n_g}\operatorname{diag}\Bigg\{ \sum_{i=1}^{n} \hat{z}_{ig} \Big[& b_{ig}(\mathbf{x}_i - \hat{\boldsymbol{\mu}}_g)(\mathbf{x}_i - \hat{\boldsymbol{\mu}}_g)' - 2\hat{\boldsymbol{\alpha}}_g(\mathbf{x}_i - \hat{\boldsymbol{\mu}}_g)' + a_{ig}\hat{\boldsymbol{\alpha}}_g\hat{\boldsymbol{\alpha}}_g' \\
& + 2\hat{\boldsymbol{\alpha}}_g\mathbf{E}_{1ig}'(\hat{\boldsymbol{\Lambda}}_g^{\text{new}})' - 2(\mathbf{x}_i - \hat{\boldsymbol{\mu}}_g)\mathbf{E}_{2ig}'(\hat{\boldsymbol{\Lambda}}_g^{\text{new}})' + \hat{\boldsymbol{\Lambda}}_g^{\text{new}}\mathbf{E}_{3ig}(\hat{\boldsymbol{\Lambda}}_g^{\text{new}})' \Big] \Bigg\}.
\end{aligned}
$$
$$\tag{6.37}$$

The AECM algorithm for the MGHFA model can now be summarized.

AECM Algorithm for the MGHFA Model

initialize $\hat{\mathbf{z}}_{ig}$

initialize $\hat{\pi}_g, \hat{\boldsymbol{\mu}}_g, \hat{\boldsymbol{\alpha}}_g, \hat{\lambda}_g, \hat{\omega}_g, \hat{\boldsymbol{\Lambda}}_g, \hat{\boldsymbol{\Psi}}_g$

while convergence criterion not met

    update $\hat{z}_{ig}, a_{ig}, b_{ig}, c_{ig}$

    update $\hat{\pi}_g, \hat{\boldsymbol{\mu}}_g, \hat{\boldsymbol{\alpha}}_g$

    update $\hat{\lambda}_g$

    update $\hat{\omega}_g$

    update $\hat{z}_{ig}, a_{ig}, b_{ig}, c_{ig}, \mathbf{E}_{1ig}, \mathbf{E}_{2ig}, \mathbf{E}_{3ig}$

    update $\hat{\boldsymbol{\Lambda}}_g, \hat{\boldsymbol{\Psi}}_g$

    check convergence criterion

end while

6.5.3 Analogy with the Gaussian Solution

To see the analogy with the solutions for the mixture of factor analyzers model (cf. Chapter 3), start by letting

$$\mathbf{S}_g = \frac{1}{n_g} \sum_{i=1}^{n} \hat{z}_{ig}[b_{ig}(\mathbf{x}_i - \hat{\boldsymbol{\mu}}_g)(\mathbf{x}_i - \hat{\boldsymbol{\mu}}_g)' - (\mathbf{x}_i - \hat{\boldsymbol{\mu}}_g)\hat{\boldsymbol{\alpha}}'_g$$
$$- \hat{\boldsymbol{\alpha}}_g(\mathbf{x}_i - \hat{\boldsymbol{\mu}}_g)' + a_{ig}\hat{\boldsymbol{\alpha}}_g\hat{\boldsymbol{\alpha}}'_g]$$

and

$$\boldsymbol{\Theta}_g = \frac{1}{n_g} \sum_{i=1}^{n} \hat{z}_{ig}\mathbf{E}_{3ig}$$
$$= \frac{1}{n_g} \sum_{i=1}^{n} \hat{z}_{ig}\{b_{ig}[\mathbf{I}_q - \hat{\boldsymbol{\beta}}_g\hat{\boldsymbol{\Lambda}}_g + \hat{\boldsymbol{\beta}}_g(\mathbf{x}_i - \hat{\boldsymbol{\mu}}_g)(\mathbf{x}_i - \hat{\boldsymbol{\mu}}_g)'\hat{\boldsymbol{\beta}}'_g]$$
$$- \hat{\boldsymbol{\beta}}_g[(\mathbf{x}_i - \hat{\boldsymbol{\mu}}_g)\hat{\boldsymbol{\alpha}}'_g + \hat{\boldsymbol{\alpha}}_g(\mathbf{x}_i - \hat{\boldsymbol{\mu}}_g)']\hat{\boldsymbol{\beta}}'_g + a_{ig}\hat{\boldsymbol{\beta}}_g\hat{\boldsymbol{\alpha}}_g\hat{\boldsymbol{\alpha}}'_g\hat{\boldsymbol{\beta}}'_g\}$$
$$= \bar{b}_g[\mathbf{I}_q - \hat{\boldsymbol{\beta}}_g\hat{\boldsymbol{\Lambda}}_g] + \hat{\boldsymbol{\beta}}_g\mathbf{S}_g\hat{\boldsymbol{\beta}}'_g.$$

From (6.36), the update for $\hat{\boldsymbol{\Lambda}}_g$ can be written

$$\hat{\boldsymbol{\Lambda}}_g^{\text{new}} = \left\{\frac{1}{n_g}\sum_{i=1}^{n}\hat{z}_{ig}\left[(\mathbf{x}_i - \hat{\boldsymbol{\mu}}_g)\mathbf{E}'_{2ig} - \hat{\boldsymbol{\alpha}}_g\mathbf{E}'_{1ig}\right]\right\}\left\{\frac{1}{n_g}\sum_{i=1}^{n}\hat{z}_{ig}\mathbf{E}_{3ig}\right\}^{-1}$$
$$= \frac{1}{n_g}\sum_{i=1}^{n}\hat{z}_{ig}\left\{(\mathbf{x}_i - \hat{\boldsymbol{\mu}}_g)[b_{ig}(\mathbf{x}_i - \hat{\boldsymbol{\mu}}_g)' - \hat{\boldsymbol{\alpha}}'_g] - \hat{\boldsymbol{\alpha}}_g[(\mathbf{x}_i - \hat{\boldsymbol{\mu}}_g)' - a_{ig}\hat{\boldsymbol{\alpha}}'_g]\right\}\hat{\boldsymbol{\beta}}'_g\boldsymbol{\Theta}_g^{-1}$$
$$= \mathbf{S}_g\hat{\boldsymbol{\beta}}'_g\boldsymbol{\Theta}_g^{-1}.$$

From (6.37), the update for $\hat{\boldsymbol{\Psi}}_g$ is given by

$$
\begin{aligned}
\hat{\boldsymbol{\Psi}}_g^{\text{new}} = \text{diag}\bigg\{ \mathbf{S}_g &+ \frac{1}{n_g} \sum_{i=1}^{n} \hat{z}_{ig} \big\{ 2\hat{\boldsymbol{\alpha}}_g [(\mathbf{x}_i - \hat{\boldsymbol{\mu}}_g)' - a_{ig}\hat{\boldsymbol{\alpha}}_g'] \\
&- 2(\mathbf{x}_i - \hat{\boldsymbol{\mu}}_g)[b_{ig}(\mathbf{x}_i - \hat{\boldsymbol{\mu}}_g)' - \hat{\boldsymbol{\alpha}}_g'] \big\} \hat{\boldsymbol{\beta}}_g'(\hat{\boldsymbol{\Lambda}}_g^{\text{new}})' + \hat{\boldsymbol{\Lambda}}_g^{\text{new}} \boldsymbol{\Theta}_g (\hat{\boldsymbol{\Lambda}}_g^{\text{new}})' \bigg\} \\
= \text{diag}\big\{ \mathbf{S}_g &- 2\mathbf{S}_g \hat{\boldsymbol{\beta}}_g'(\hat{\boldsymbol{\Lambda}}_g^{\text{new}})' + \hat{\boldsymbol{\Lambda}}_g^{\text{new}} \boldsymbol{\Theta}_g \big(\boldsymbol{\Theta}_g^{-1}\hat{\boldsymbol{\beta}}_g \mathbf{S}_g\big)\big\} \\
= \text{diag}\big\{ \mathbf{S}_g &- \hat{\boldsymbol{\Lambda}}_g^{\text{new}} \hat{\boldsymbol{\beta}}_g \mathbf{S}_g \big\}.
\end{aligned}
$$

6.6 Illustrations

6.6.1 Old Faithful Data

The famous Old Faithful geyser data give measurements on the waiting time between and the duration of 272 eruptions of the Old Faithful geyser, which is located in Yellowstone National Park. These data are available as `faithful` in R, and they have been used many times to illustrate approaches to analyzing skewed data (e.g., Ali et al., 2010; Vrbik and McNicholas, 2012). Mixtures of SAL distributions and mixtures of generalized hyperbolic distributions are fitted to the geyser data for $G = 1, 2, 3$. In both cases, a $G = 2$ component model is selected and the associated contour plots present a nice contrast (Figure 6.5).

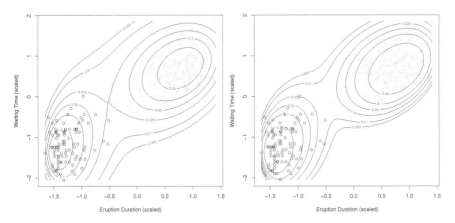

Figure 6.5

Scatter plots for the Old Faithful data, with contours from a fitted $G = 2$ component SAL mixture (left) and contours from a $G = 2$ component MGHD (right), where plotting symbols and shading represent predicted classifications in each case.

Specifically, while both approaches give the same classification results, the MGHD gives a better fit to the data. The component concentration parameters enable the generalized hyperbolic components to concentrate around the modes; on the other hand, the SAL distribution only parameterizes location, scale, and skewness. The precise impact of the component index parameters is difficult to see here. As Browne and McNicholas (2015) point out, the fact that the component index parameters are free to vary "engenders a flexibility not present in other non-elliptical mixture approaches to clustering". They support this claim via log-density plots and comparative analyses.

6.6.2 Yeast Data

The yeast data contain cellular localization sites of 1,484 proteins, and they are available from the UCI machine learning repository. The development of these data, as well as classification results using a "rule-based expert system", are discussed by Nakai and Kanehisa (1991, 1992). Following Franczak et al. (2014) and Murray et al. (2014a), consider three variables: McGeoch's method for signal sequence recognition (MCG), the score of the ALOM membrane spanning region prediction program (ALM), and the score of discriminant analysis of the amino acid content of vacuolar and extracellular proteins (VAC). Two localization sites, CYT (cytosolic or cytoskeletal) and ME3 (membrane protein, no N-terminal signal), can be considered the "true" classes. Inspection of a pairs plot of these data (Figure 6.6) demonstrates it is a difficult clustering problem.

First, consider clustering. SAL and generalized hyperbolic mixture models are fitted to these data for $G = 1, \ldots, 5$ components. The chosen SAL mixture model has $G = 2$ components, as does the chosen MGHD. In both cases, the associated MAP classification results are very good, with the generalized hyperbolic mixture giving slightly better performance (Table 6.1).

Table 6.1
Cross-tabulations of localization sites against predicted classifications for the chosen SAL and generalized hyperbolic mixture models, respectively, for the yeast data.

	SAL Mixture		MGHD	
	A	B	A	B
CYT	448	15	449	13
ME3	14	149	14	150

For illustration, a Gaussian mixture model is also fitted for $G = 1, \ldots, 5$. A $G = 3$ component model is selected and the associated classification performance (Table 6.2; ARI = 0.57) is notably worse than the chosen SAL mixture (ARI = 0.81) and the chosen generalized hyperbolic mixture (ARI = 0.83).

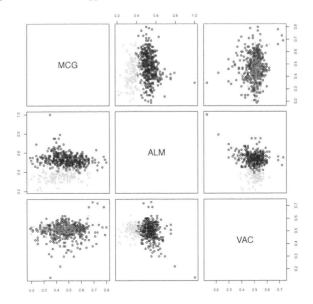

Figure 6.6
Pairs plot for the yeast data, where plotting symbols and shading correspond
to location sites (CYT and ME3).

Moreover, merging Gaussian components does not improve the situation;
in fact, the most favourable merging leads to an inferior classification result (ARI = 0.47). Franczak et al. (2014) suggest that, for completeness, a
$G = 2$ component Gaussian mixture model should be fitted to these data; as
they show, the associated classification performance is very poor (Table 6.2;
ARI $= -0.09 < 0$).

Table 6.2
Cross-tabulations of localization sites against predicted classifications for $G = 3$ and $G = 2$ component Gaussian mixture models, respectively, for the yeast
data.

| | $G = 3$ | | | $G = 2$ | |
	A	B	C	A	B
CYT	379	12	72	106	357
ME3	13	11	139	1	162

Because these data present a rather difficult clustering problem, it is of
interest to apply them for model-based classification. Franczak et al. (2014)
compare model-based classification within the SAL and Gaussian mixture
modelling frameworks, by analyzing the yeast data with 70% of the observations treated as labelled. Specifically, they consider 25 different random 70/30

partitions of the data. These results are considered here as well as the analogous results for MGHD. The aggregate classification results for the three methods (Table 6.3) reveal that the generalized hyperbolic (ARI = 0.86) and SAL (ARI = 0.86) mixtures give far better classification performance than the Gaussian mixture (ARI = −0.080).

Table 6.3
Cross-tabulation of true versus (aggregate) predicted classification results for generalized hyperbolic, SAL, and Gaussian mixture models, respectively, for the yeast data for 25 runs with 70% of the labels taken as known.

	MGHD		SAL mixture		Gaussian mixture	
	A	B	A	B	A	B
CYT	3406	70	3403	71	964	2502
ME3	88	1136	87	1139	29	1205

The poor performance of the Gaussian mixture modelling approach underlines the fact that a Gaussian mixture model cannot be relied upon for semi-supervised classification when the data have asymmetric clusters, even when most (here, 70%) of the samples are labelled.

6.6.3 Italian Wine Data

The Italian wine data were used to illustrate the mixture of factor analyzers model as well as extensions thereof — perhaps most notably the PGMM family (cf. Chapter 3). Recall that these data arise from physical and chemical analysis of wines grown in the same region in Italy but derived from three different cultivars: Barolo, Grignolino, and Barbera. In all, 178 samples of 27 physical and chemical measurements are available in the pgmm package. Tortora et al. (2015b) use these data to illustrate the performance of the MGHFA model. They fix $G = 3$ and choose the number of factors q based on the BIC. This process is repeated for the mixture of factor analyzers model. For the MGHFA model, a $q = 2$ factor model is selected and the associated MAP classifications correspond quite well to the cultivars (Table 6.4; ARI = 0.80). For the mixture of factor analyzers model, a $q = 3$ factor model is selected and the associated classification performance is very similar to that for the MGHFA model (Table 6.4; ARI = 0.79).

6.6.4 Liver Data

Chen et al. (2002) use gene expression microarray data to compare patients with hepatocellular carcinoma (HCC) to people with healthy (non-tumour) liver tissue. In all, there are 179 tissue samples: 104 from patients with HCC and 75 non-tumour samples. For each tissue sample, there are 85 gene expres-

Table 6.4

Cross-tabulation of cultivar versus predicted classifications for model-based clustering of the Italian wine data using the MGHFA and mixture of factor analyzers (MFA) models, respectively.

	MGHFA			MFA		
	1	2	3	1	2	3
Barolo	50	0	0	59	0	0
Grignolino	10	60	1	4	58	9
Barbera	0	1	47	0	0	48

sions. These data serve as a good example for the MGHFA model; in fact, they were used by Tortora et al. (2015b) for this very purpose. Following the same procedure as in Section 6.6.3, the number of components is fixed at $G = 2$ and the BIC selects $q = 2$ latent factors. The MGHFA model gives very good classification performance (Table 6.5; ARI = 0.672); however, the mixture of factor analyzers model gives classifications that are no better than one would expect from guessing (Table 6.5; ARI = −0.01).

Table 6.5

Cross-tabulation of tissue type versus predicted classifications for model-based clustering of the liver data using the MGHFA and mixture of factor analyzers (MFA) models, respectively.

	MGHFA		MFA	
	1	2	1	2
HCC	97	7	13	91
Non-tumour	9	66	8	67

6.7   A Note on Normal Variance-Mean Mixtures

The SAL and generalized hyperbolic distributions arise as normal variance-mean mixtures, cf. (6.18). Formally, it should be noted that the distribution of a p-dimensional random variable $\mathbf{X}$ is said to be a normal variance-mean mixture if its density can be written in the form

$$f(\mathbf{x} \mid \boldsymbol{\mu}, \boldsymbol{\Sigma}, \boldsymbol{\alpha}, \boldsymbol{\theta}) = \int_0^\infty \phi_p\left(\mathbf{x} \mid \boldsymbol{\mu} + w\boldsymbol{\alpha}, w\boldsymbol{\Sigma}\right) h\left(w \mid \boldsymbol{\theta}\right) dw, \qquad (6.38)$$

where $\phi_p\left(\mathbf{x} \mid \boldsymbol{\mu}+w\boldsymbol{\alpha}, w\boldsymbol{\Sigma}\right)$ is the density of a p-dimensional Gaussian distribution with mean $\boldsymbol{\mu}+w\boldsymbol{\alpha}$ and covariance matrix $w\boldsymbol{\Sigma}$, and $h\left(w \mid \boldsymbol{\theta}\right)$ is the density of a univariate random variable $W > 0$ (cf. Barndorff–Nielsen et al., 1982; Gneiting, 1997). Now $h\left(w \mid \boldsymbol{\theta}\right)$ is a weight function that can take on many forms. For the SAL distribution, $h\left(w \mid \boldsymbol{\theta}\right)$ is the density of an exponential random variable with rate 1 and, for the generalized hyperbolic distribution, it is the density of a GIG distribution. Apart from those mentioned in this chapter, there are other possibilities for $h\left(w \mid \boldsymbol{\theta}\right)$, some of which lead to density representations for well-known non-Gaussian distributions. For example, if $h\left(w \mid \boldsymbol{\theta}\right)$ is the density of an inverse-gamma random variable with parameters $(\nu/2, \nu/2)$, then (6.38) is a representation of the skew-t distribution with ν degrees of freedom (cf. Demarta and McNeil, 2005; Murray et al., 2014a). Note that the representation of the t-distribution discussed in Section 5.1 is equivalent to (6.38) with the same $h\left(w \mid \boldsymbol{\theta}\right)$, i.e., inverse-gamma$(\nu/2, \nu/2)$, and $\boldsymbol{\alpha} = \mathbf{0}$. If $h\left(w \mid \boldsymbol{\theta}\right)$ is the density of a gamma distribution, then (6.38) is a representation of the variance-gamma distribution (cf. McNicholas et al., 2014). Further details on, and examples of, normal variance-mean mixtures are given by Barndorff–Nielsen (1978), Kotz et al. (2001), and Kotz and Nadarajah (2004), amongst others.

6.8 Comments

There has been a plethora of work on clustering using non-elliptical distributions beyond the MGHD and special and/or limiting cases thereof. The decision to restrict the contents of this chapter to the mixture of SAL distributions and the MGHD is partly due to the fact that multiple scaled analogues thereof are discussed in Chapter 7. Of course, the SAL mixture is attractive as a first departure from mixtures with symmetric components because it only parameterizes location, scale, and skewness. The MGHD, also parameterizing concentration (as well as having an index parameter), is a natural extension.

   Within this monograph, this chapter might have been replaced by one that began with the skew-normal distribution and proceeded to the skew-t distribution. Of course, the smooth progression to the mixtures of multiple scaled distributions, in Chapter 7, would have been lost. Furthermore, there would have been the matter of choosing which formulations of the skew-normal and skew-t distributions ought to be elaborated upon. In fact, even the discussion about this matter would have somewhat obfuscated the material. Mixtures of skew-normal distributions and mixtures of skew-t distributions have garnered significant attention within the model-based clustering and classification literature and, notwithstanding the aforementioned issues, they deserve some mention herein.

   A good starting point is the (multivariate) skew-normal distribution proposed by Azzalini and Valle (1996) and subsequently examined further by Azzalini and Capitanio (1999) and others. This formulation has the characterization

$$\mathbf{X} = \boldsymbol{\mu} + \boldsymbol{\gamma}|Z_0| + \mathbf{Z}_1, \tag{6.39}$$

where $\mathbf{X}$ is a p-dimensional skew-normal random variable, $\boldsymbol{\mu}$ is a location parameter,

$$\begin{bmatrix} Z_0 \\ \mathbf{Z}_1 \end{bmatrix} \sim \mathrm{N}_{1+p} \left(\begin{bmatrix} 0 \\ \mathbf{0}_p \end{bmatrix}, \begin{bmatrix} 1 & \mathbf{0}'_p \\ \mathbf{0}_p & \boldsymbol{\Sigma} - \boldsymbol{\gamma}\boldsymbol{\gamma}' \end{bmatrix} \right),$$

$\boldsymbol{\Sigma}_g$ is the scale matrix, and $\boldsymbol{\gamma}$ is the skewness parameter. The formulation associated with (6.39) can be referred to as the classical skew-normal distribution — see Azzalini et al. (2016) for a discussion on nomenclature for this and another formulation of the skew-normal distribution.

Another formulation of the skew-normal distribution is given by Sahu et al. (2003), and has the characterization

$$\mathbf{X} = \boldsymbol{\mu} + \boldsymbol{\Gamma}|\mathbf{Z}_0| + \mathbf{Z}_1, \tag{6.40}$$

where $\mathbf{X}$ is a p-dimensional skew-normal random variable, $\boldsymbol{\mu}$ is a location parameter,

$$\begin{bmatrix} \mathbf{Z}_0 \\ \mathbf{Z}_1 \end{bmatrix} \sim \mathrm{N}_{2p} \left(\begin{bmatrix} \mathbf{0}_p \\ \mathbf{0}_p \end{bmatrix}, \begin{bmatrix} \mathbf{I}_p & \mathbf{0}_p \\ \mathbf{0}_p & \boldsymbol{\Sigma} - \boldsymbol{\Gamma}\boldsymbol{\Gamma}' \end{bmatrix} \right),$$

and $\boldsymbol{\Gamma}$ is a $p \times q$ skewness matrix. Note that, in effect, $|\mathbf{Z}_0|$ is a half-normal random variable. The formulation associated with (6.40) can be referred to as the SDB skew-normal distribution, based on the last names of the relevant authors (i.e., Sahu, Dey, and Branco; cf. Azzalini et al., 2016). Extensive details on skew-normal distributions are given by Azzalini and Capitanio (2014).

The classical and SDB formulations of the skew-normal distribution only coincide for $p = 1$. Furthermore, for $p > 1$, the two formulations differ in more than just their parameterization — this is true despite the fact that they have the same number of free parameters; details are given by Azzalini et al. (2016). These points also hold true for the skew-t analogues of these skew-normal distributions. Branco and Dey (2001) and Azzalini and Capitanio (2003) introduce a classical skew-t distribution, and Sahu et al. (2003) consider an SDB skew-t distribution. Most importantly in the context of the present monograph, these skew-normal and skew-t formulations have been used for model-based clustering and classification, including work by Lin (2009, 2010), Vrbik and McNicholas (2012, 2014), and Lee and McLachlan (2013a,b, 2014). Vrbik and McNicholas (2014) introduce skew-normal and skew-t analogues of the GPCM family and show that they can give superior clustering and classification performance when compared with their Gaussian counterparts. Murray et al. (2013) develop a mixture of skew-t factor analyzers model using the SDB formulation, and Lin et al. (2016) discuss a mixture of skew-normal factor analyzers model using the classical formulation.

Finally, it is worth reiterating that a formulation of the skew-t distribution arises from the generalized hyperbolic distribution. This formulation is used by Murray et al. (2014a) and can be obtained via

$$\mathbf{X} = \boldsymbol{\mu} + W\boldsymbol{\alpha} + \sqrt{W}\mathbf{V}, \tag{6.41}$$

where $\boldsymbol{\mu}$ is the location parameter, $\boldsymbol{\alpha}$ is the skewness parameter, $\mathbf{V} \sim \mathrm{N}(\mathbf{0}, \boldsymbol{\Sigma})$ and $W \sim \mathrm{IG}(\nu/2, \nu/2)$, where $\mathrm{IG}(\cdot)$ denotes the inverse-gamma distribution. In fact, Murray et al. (2014a) develop a skew-t factor analysis model that arises by setting

$$\mathbf{V} = \boldsymbol{\Lambda}\mathbf{U} + \boldsymbol{\varepsilon}, \tag{6.42}$$

where $\boldsymbol{\Lambda}$ is a matrix of factor loadings, $\mathbf{U} \sim \mathrm{N}(\mathbf{0}, \mathbf{I}_q)$ are the latent factors, and $\boldsymbol{\varepsilon} \sim \mathrm{N}(\mathbf{0}, \boldsymbol{\Psi})$ with $\boldsymbol{\Psi} = \mathrm{diag}(\psi_1, \psi_2, \ldots, \psi_p)$, and the usual assumptions hold. From (6.41) and (6.42), a skew-t factor analysis model is given by

$$\mathbf{X} = \boldsymbol{\mu} + W\boldsymbol{\alpha} + \sqrt{W}(\boldsymbol{\Lambda}\mathbf{U} + \boldsymbol{\varepsilon}). \tag{6.43}$$

Proceeding in a similar fashion to Section 6.5, a mixture of these skew-t factor analyzers is implemented by Murray et al. (2014a) and further details are available therein. Interestingly, Murray et al. (2014b) use the same formulation, i.e., what is sometimes called the generalized hyperbolic skew-t distribution, to obtain a mixture of common skew-t factor analyzers.

7

Mixtures of Multiple Scaled Distributions

7.1 Overview

After considering mixtures where the component densities parameterize location, scale, concentration, skewness, and index (Chapter 6), the next step in increasing modelling flexibility is allowing multiple scaled components. An argument as to why this might be desirable is best made by looking at a specific example of data for which multiple scaled components are needed. Consider the `bankruptcy` data from the `MixGHD` package. These data contain the ratio of retained earnings to total assets as well as the ratio of earnings before interest and taxes to total assets for 66 American firms (Figure 7.1).

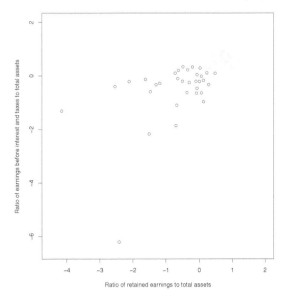

Figure 7.1
Scatter plot for the (scaled) bankruptcy data, where plotting symbol and shading represent status, i.e., filed for bankruptcy or financially sound.

Each firm has either filed for bankruptcy or is financially sound, and the data were first reported by Altman (1968). The scatter plot in Figure 7.1

reveals that the bankrupts (denoted ∘) disperse in two directions: one corresponding to both ratios being in similarly poor health, and the other indicating a healthier ratio of earnings before interest and taxes to total assets.

7.2   Mixture of Multiple Scaled t-Distributions

Recall that the distribution of a p-dimensional random variable $\mathbf{X}$ is said to be a normal variance-mean mixture if its density can be written in the form

$$f(\mathbf{x} \mid \boldsymbol{\mu}, \boldsymbol{\Sigma}, \boldsymbol{\alpha}, \boldsymbol{\theta}) = \int_0^\infty \phi_p\left(\mathbf{x} \mid \boldsymbol{\mu} + w\boldsymbol{\alpha}, w\boldsymbol{\Sigma}\right) h\left(w \mid \boldsymbol{\theta}\right) dw, \qquad (7.1)$$

where $\phi_p\left(\mathbf{x} \mid \boldsymbol{\mu} + w\boldsymbol{\alpha}, w\boldsymbol{\Sigma}\right)$ is the density of a p-dimensional Gaussian distribution with mean $\boldsymbol{\mu} + w\boldsymbol{\alpha}$ and covariance matrix $w\boldsymbol{\Sigma}$, and $h\left(w \mid \boldsymbol{\theta}\right)$ is the probability density function of a univariate random variable $W > 0$. As discussed in Section 6.7, the density $h\left(w \mid \boldsymbol{\theta}\right)$ is a weight function that can take on many forms leading, in several cases, to well-known distributions. Setting $\boldsymbol{\alpha} = \mathbf{0}$ and

$$h\left(w \mid \nu/2, \nu/2\right) = w^{\nu/2-1} \Gamma\left(\nu/2\right)^{-1} \exp\{-\nu w/2\}(\nu/2)^{\nu/2} \qquad (7.2)$$

in (7.1), where $\Gamma(\cdot)$ is the gamma function, it follows that the density of the multivariate t-distribution with ν degrees of freedom can be written

$$f_t(\mathbf{x} \mid \boldsymbol{\mu}, \boldsymbol{\Sigma}, \nu) = \int_0^\infty \phi_p\left(\mathbf{x} \mid \boldsymbol{\mu}, \boldsymbol{\Sigma}/w\right) h\left(w \mid \nu/2, \nu/2\right) dw$$

$$= \frac{\Gamma\left([\nu+p]/2\right) |\boldsymbol{\Sigma}|^{-1/2}}{(\pi\nu)^{p/2} \Gamma\left(\nu/2\right) \left[1 + \delta(\mathbf{x}, \boldsymbol{\mu} \mid \boldsymbol{\Sigma})/\nu\right]^{(\nu+p)/2}}. \qquad (7.3)$$

Note that (7.2) is the gamma density, for $w > 0$, and the formulation of the t-distribution in (7.3) is the same as in (5.3).

Forbes and Wraith (2014) show that a multi-dimensional weight variable $\boldsymbol{\Delta}_{\mathbf{W}} = \text{diag}\left(W_1, \ldots, W_p\right)$ can be incorporated into (7.1) via an eigen-decomposition of the symmetric positive-definite matrix $\boldsymbol{\Sigma}$. Specifically, they set $\boldsymbol{\Sigma} = \boldsymbol{\Gamma}\boldsymbol{\Phi}\boldsymbol{\Gamma}'$, where $\boldsymbol{\Gamma}$ is a $p \times p$ matrix of eigenvectors and $\boldsymbol{\Phi}$ is a $p \times p$ diagonal matrix containing the eigenvalues of $\boldsymbol{\Sigma}$. It follows that the density of $\mathbf{X}$ becomes

$$f\left(\mathbf{x} \mid \boldsymbol{\mu}, \boldsymbol{\Gamma}, \boldsymbol{\Phi}, \boldsymbol{\alpha}, \boldsymbol{\theta}\right) =$$

$$\int_0^\infty \cdots \int_0^\infty \phi_p\left(\mathbf{x} \mid \boldsymbol{\mu} + \boldsymbol{\Delta}_{\mathbf{W}}\boldsymbol{\alpha}, \boldsymbol{\Gamma}\boldsymbol{\Delta}_{\mathbf{W}}\boldsymbol{\Phi}\boldsymbol{\Gamma}'\right) h_{\mathbf{W}}\left(w_1, \ldots, w_p \mid \boldsymbol{\theta}\right) dw_1 \ldots dw_p,$$

$$(7.4)$$

where

$$h_{\mathbf{W}}\left(w_1, \ldots, w_p \mid \boldsymbol{\theta}\right) = h\left(w_1 \mid \boldsymbol{\theta}_1\right) \times \cdots \times h\left(w_p \mid \boldsymbol{\theta}_p\right)$$

is a p-dimensional density such that the random variables $W_1, \ldots, W_p$ are independent, i.e., the weights are independent. The density given in (7.4) adds flexibility to normal variance-mean mixtures because the parameters $\boldsymbol{\theta}_1, \ldots, \boldsymbol{\theta}_p$ are free to vary in each dimension. Using the density in (7.4), Forbes and Wraith (2014) derive the density of a multiple scaled multivariate-t distribution, Tortora et al. (2014) introduce a multiple scaled generalized hyperbolic distribution (see Section 7.4), Wraith and Forbes (2014) derive a multiple scaled normal-inverse Gaussian distribution, and Franczak et al. (2015) develop a multiple scaled SAL distribution.

Setting $\boldsymbol{\Sigma} = \boldsymbol{\Gamma}\boldsymbol{\Phi}\boldsymbol{\Gamma}'$, it follows from (7.4) that the density of a multiple scaled analogue of (7.3) can be written

$$f_{t\text{MS}}(\mathbf{x} \mid \boldsymbol{\mu}, \boldsymbol{\Gamma}, \boldsymbol{\Phi}, \boldsymbol{\nu}) =$$
$$\int_0^\infty \cdots \int_0^\infty \phi_p\left(\mathbf{x} \mid \boldsymbol{\mu}, \boldsymbol{\Gamma}\boldsymbol{\Phi}\boldsymbol{\Delta_W}\boldsymbol{\Gamma}'\right) h_{\mathbf{W}}(w_1, \ldots, w_p \mid \boldsymbol{\nu}) \, dw_1 \ldots dw_p, \tag{7.5}$$

where $\boldsymbol{\Delta_W} = \text{diag}\left(w_1^{-1}, \ldots, w_p^{-1}\right)$ and the weight function

$$h_{\mathbf{W}}(w_1, \ldots, w_p \mid \boldsymbol{\nu}) = h\left(w_1 \mid \nu_1/2, \nu_1/2\right) \times \cdots \times h\left(w_p \mid \nu_p/2, \nu_p/2\right)$$

is a p-dimensional gamma density, where $h\left(w_j \mid \nu_j/2, \nu_j/2\right)$ is given by (7.2). Note that the scaled Gaussian density in (7.5) can be written

$$\phi_p\left(\mathbf{x} \mid \boldsymbol{\mu}, \boldsymbol{\Gamma}\boldsymbol{\Phi}\boldsymbol{\Delta_W}\boldsymbol{\Gamma}'\right) = \prod_{j=1}^p \phi_1\left([\boldsymbol{\Gamma}'(\mathbf{x} - \boldsymbol{\mu})]_j \mid 0, \Phi_j w_j^{-1}\right), \tag{7.6}$$

where $\phi_1\left([\boldsymbol{\Gamma}'(\mathbf{x} - \boldsymbol{\mu})]_j \mid 0, \Phi_j w_j^{-1}\right)$ is the density of a univariate Gaussian distribution with mean 0 and variance $\Phi_j w_j^{-1}$, $[\boldsymbol{\Gamma}'(\mathbf{x} - \boldsymbol{\mu})]_j$ is the jth element of $\boldsymbol{\Gamma}'(\mathbf{x} - \boldsymbol{\mu})$, and Φ_j is the jth eigenvalue of $\boldsymbol{\Phi}$. It follows that (7.5) can be written

$$f_{t\text{MS}}(\mathbf{x} \mid \boldsymbol{\mu}, \boldsymbol{\Gamma}, \boldsymbol{\Phi}, \boldsymbol{\nu}) =$$
$$\prod_{j=1}^p \int_0^\infty \phi_1\left([\boldsymbol{\Gamma}'(\mathbf{x} - \boldsymbol{\mu})]_j \mid 0, \Phi_j w_j^{-1}\right) h\left(w_j \mid \nu_j/2, \nu_j/2\right) dw_j. \tag{7.7}$$

Solving the integral in (7.7) gives the density of a multiple scaled multivariate-t distribution,

$$f_{t\text{MS}}(\mathbf{x} \mid \boldsymbol{\mu}, \boldsymbol{\Gamma}, \boldsymbol{\Phi}, \boldsymbol{\nu}) = \prod_{j=1}^p \frac{\Gamma([\nu_j + 1]/2)}{\Gamma(\nu_j/2)(\Phi_j \nu_j \pi)^{1/2}} \left[1 + \frac{[\boldsymbol{\Gamma}'(\mathbf{x} - \boldsymbol{\mu})]_j^2}{\Phi_j \nu_j}\right]^{-(\nu_j + 1)/2}, \tag{7.8}$$

where Φ_j is the jth eigenvalue of $\boldsymbol{\Phi}$, $\boldsymbol{\Gamma}$ is a matrix of eigenvectors, $\boldsymbol{\mu}$ is a location parameter, and $[\boldsymbol{\Gamma}'(\mathbf{x} - \boldsymbol{\mu})]_j^2/\Phi_j$ can be regarded as the squared Mahalanobis distance between $\mathbf{x}$ and $\boldsymbol{\mu}$.

The main difference between the traditional multivariate-t density given in (7.3) and the multiple scaled multivariate-t density given in (7.8) is that the degrees of freedom can now be parameterized separately in each dimension j. Therefore, unlike the standard multivariate-t distribution, the multiple scaled density in (7.8) can account for different tail weight in each dimension (cf. Forbes and Wraith, 2014).

7.3   Mixture of Multiple Scaled SAL Distributions

Recall that a SAL distribution can be written as a normal variance-mean mixture where the univariate density is that of an exponentially distributed random variable with rate 1, i.e., $h(w) = e^{-w}$ for $w > 0$. It follows that the SAL density can be written

$$f(\mathbf{x} \mid \boldsymbol{\mu}, \boldsymbol{\Sigma}, \boldsymbol{\alpha}) = \int_0^\infty \phi_p\left(\mathbf{x} \mid \boldsymbol{\mu} + w\boldsymbol{\alpha}, w\boldsymbol{\Sigma}\right) h\left(w\right) dw. \qquad (7.9)$$

Franczak et al. (2015) use (7.4) and (7.9) to write the density of a multiple scaled SAL (MSSAL) distribution as

$$f_{\mathrm{MSS}}(\mathbf{x} \mid \boldsymbol{\alpha}, \boldsymbol{\Gamma}, \boldsymbol{\Phi}, \boldsymbol{\mu}) =$$
$$\int_0^\infty \cdots \int_0^\infty \phi_p\left(\mathbf{x} \mid \boldsymbol{\mu} + \boldsymbol{\Delta}_{\mathbf{w}}\boldsymbol{\alpha}, \boldsymbol{\Gamma}\boldsymbol{\Phi}\boldsymbol{\Delta}_{\mathbf{w}}\boldsymbol{\Gamma}'\right) h_{\mathbf{W}}\left(w_1, \ldots, w_p\right) dw_1 \ldots dw_p,$$

where $h_{\mathbf{W}}(w_1, \ldots, w_p) = h(w_1) \times \cdots \times h(w_p)$.

To simplify parameter estimation, Franczak et al. (2015) let $\boldsymbol{\Delta}_{\mathbf{w}}\boldsymbol{\alpha} = \boldsymbol{\Omega}\boldsymbol{\beta}$, where $\boldsymbol{\beta} \in \mathbb{R}^p$ is a transformed skewness parameter and $\boldsymbol{\Omega} = \boldsymbol{\Gamma}\boldsymbol{\Phi}\boldsymbol{\Delta}_{\mathbf{w}}\boldsymbol{\Gamma}'$. From (7.6), it follows that

$$f_{\mathrm{MSS}}(\mathbf{x} \mid \boldsymbol{\beta}, \boldsymbol{\Gamma}, \boldsymbol{\Phi}, \boldsymbol{\mu}) = \prod_{j=1}^p \int_0^\infty \phi_1\left(\left[\boldsymbol{\Gamma}'(\mathbf{x} - \boldsymbol{\mu} - \boldsymbol{\Phi}\boldsymbol{\Delta}_{\mathbf{w}}\boldsymbol{\beta})\right]_j \mid 0, \Phi_j w_j\right) h\left(w_j\right) dw_j$$
$$= \prod_{j=1}^p \frac{1}{\gamma_j} \exp\left\{\frac{-\left|\left[\boldsymbol{\Gamma}'(\mathbf{x} - \boldsymbol{\mu})\right]_j\right|}{\Phi_j}\left[\gamma_j - \left[\boldsymbol{\Phi}\boldsymbol{\Gamma}'\boldsymbol{\beta}\right]_j \operatorname{sign}\left\{\left[\boldsymbol{\Gamma}'(\mathbf{x} - \boldsymbol{\mu})\right]_j\right\}\right]\right\},$$

where $\gamma_j = \sqrt{\left[\boldsymbol{\Phi}\boldsymbol{\Gamma}'\boldsymbol{\beta}\right]_j^2 + 2\Phi_j}$.

The density of a mixture of MSSAL distributions is given by

$$f(\mathbf{x} \mid \boldsymbol{\vartheta}) = \sum_{g=1}^G \pi_g f_{\mathrm{MSS}}\left(\mathbf{x} \mid \boldsymbol{\beta}_g, \boldsymbol{\Gamma}_g, \boldsymbol{\Phi}_g, \boldsymbol{\mu}_g\right)$$

and, as Franczak et al. (2015) point out, each component of this mixture has contours that are shaped like hypercubes.

7.4 Mixture of Multiple Scaled Generalized Hyperbolic Distributions

Recall that a generalized hyperbolic distribution can be written as a normal variance-mean mixture where the univariate density is that of a random variable following a GIG distribution, i.e.,

$$h(w \mid \omega, 1, \lambda) = \frac{w^{\lambda-1}}{2K_\lambda(\omega)} \exp\left\{-\frac{\omega}{2}\left(w + \frac{1}{w}\right)\right\}, \tag{7.10}$$

for $w > 0$. It follows that the generalized hyperbolic density can be written

$$f(\mathbf{x} \mid \boldsymbol{\mu}, \boldsymbol{\Sigma}, \boldsymbol{\alpha}, \omega, \lambda) = \int_0^\infty \phi_p\left(\mathbf{x} \mid \boldsymbol{\mu} + w\boldsymbol{\alpha}, w\boldsymbol{\Sigma}\right) h(w \mid \omega, 1, \lambda) dw, \tag{7.11}$$

where $\boldsymbol{\mu}$ is the location parameter, $\boldsymbol{\alpha}$ is the skewness parameter, $\boldsymbol{\Sigma}$ is the scale matrix, λ is the index parameter, and ω is the concentration parameter; cf. Section 6.4.1.

Tortora et al. (2014, 2016) use (7.4) and (7.11) to write the density of a multiple scaled generalized hyperbolic distribution (MSGHD) as

$$f_{\text{MSGHD}}(\mathbf{x} \mid \boldsymbol{\mu}, \boldsymbol{\Gamma}, \boldsymbol{\Phi}, \boldsymbol{\alpha}, \boldsymbol{\omega}, \boldsymbol{\lambda}) =$$
$$\int_0^\infty \cdots \int_0^\infty \phi_p\left(\boldsymbol{\Gamma}'\mathbf{x} - \boldsymbol{\mu} - \boldsymbol{\Delta}_\mathbf{w}\boldsymbol{\alpha} \mid \mathbf{0}, \boldsymbol{\Delta}_\mathbf{w}\boldsymbol{\Phi}\right) \tag{7.12}$$
$$\times h_\mathbf{w}(w_1, \ldots, w_p \mid \boldsymbol{\omega}, \mathbf{1}, \boldsymbol{\lambda}) dw_1 \ldots dw_p,$$

where $\boldsymbol{\omega} = (\omega_1, \ldots, \omega_p)'$, $\boldsymbol{\lambda} = (\lambda_1, \ldots, \lambda_p)'$, $\mathbf{1}$ is a p-vector of 1s, and

$$h_\mathbf{W}(w_1, \ldots, w_p \mid \boldsymbol{\omega}, \mathbf{1}, \boldsymbol{\lambda}) = h(w_1 \mid \omega_1, 1, \lambda_1) \times \cdots \times h(w_p \mid \omega_p, 1, \lambda_p).$$

From (7.6), it follows that (7.12) can be written

$$f_{\text{MSGHD}}(\mathbf{x} \mid \boldsymbol{\mu}, \boldsymbol{\Gamma}, \boldsymbol{\Phi}, \boldsymbol{\alpha}, \boldsymbol{\omega}, \boldsymbol{\lambda})$$
$$= \prod_{j=1}^p \left\{ \left[\frac{\omega_j + \Phi_j^{-1}\left(\left[\boldsymbol{\Gamma}'\mathbf{x}\right]_j - \mu_j\right)^2}{\omega_j + \alpha_j^2\Phi_j^{-1}}\right]^{(\lambda_j - 1/2)/2} \right.$$
$$\left. \times \frac{K_{\lambda_j - 1/2}\left(\sqrt{\left[\omega_j + \alpha_j^2\Phi_j^{-1}\right]\left[\omega_j + \Phi_j^{-1}\left(\left[\boldsymbol{\Gamma}'\mathbf{x}\right]_j - \mu_j\right)^2\right]}\right)}{(2\pi)^{1/2}\Phi_j^{1/2}K_{\lambda_j}(\omega_j)\exp\left\{\left(\left[\boldsymbol{\Gamma}'\mathbf{x}\right]_j - \mu_j\right)\Phi_j^{-1}\alpha_j\right\}} \right\},$$

where $\left[\boldsymbol{\Gamma}'\mathbf{x}\right]_j$ is the jth element of the vector $\boldsymbol{\Gamma}'\mathbf{x}$, μ_j is the jth element of the location parameter $\boldsymbol{\mu}$, α_j is the jth element of the skewness parameter $\boldsymbol{\alpha}$, $\boldsymbol{\Gamma}$ is a $p \times p$ matrix of eigenvectors, Φ_j is the jth eigenvalue of the diagonal matrix $\mathbf{\Phi}$, $\boldsymbol{\omega} = (\omega_1, \ldots, \omega_p)'$ controls the concentration in each dimension, and $\boldsymbol{\lambda} = (\lambda_1, \ldots, \lambda_p)'$ is a p-dimensional index parameter. Write $\mathbf{X} \backsim \text{MSGHD}(\boldsymbol{\mu}, \mathbf{\Gamma}, \mathbf{\Phi}, \boldsymbol{\alpha}, \boldsymbol{\omega}, \boldsymbol{\lambda})$ to indicate that the random vector $\mathbf{X}$ follows an MSGHD with density $f_{\text{MSGHD}}(\mathbf{x} \mid \boldsymbol{\mu}, \mathbf{\Gamma}, \mathbf{\Phi}, \boldsymbol{\alpha}, \boldsymbol{\omega}, \boldsymbol{\lambda})$.

Tortora et al. (2014, 2016) consider a mixture of MSGHDs (MMSGHDs) with density given by

$$f(\mathbf{x} \mid \boldsymbol{\vartheta}) = \sum_{g=1}^{G} \pi_g f_{\text{MSGHD}} \left(\mathbf{x} \mid \boldsymbol{\mu}_g, \mathbf{\Gamma}_g, \mathbf{\Phi}_g, \boldsymbol{\alpha}_g, \boldsymbol{\omega}_g, \boldsymbol{\lambda}_g \right). \tag{7.13}$$

7.5   Mixture of Coalesced Generalized Hyperbolic Distributions

As Tortora et al. (2016) point out, the generalized hyperbolic distribution is not a special or limiting case of the MSGHD under any parameterization with $p > 1$. Motivated by this, they propose a coalesced generalized hyperbolic distribution (CGHD) that contains both the generalized hyperbolic distribution and MSGHD as limiting cases. The CGHD arises through the introduction of a random vector

$$\mathbf{R} = U\mathbf{X} + (1 - U)\mathbf{S}, \tag{7.14}$$

where $\mathbf{X} = \mathbf{\Gamma Y}$, $\mathbf{Y} \backsim \text{GHD}(\boldsymbol{\mu}, \mathbf{\Sigma}, \boldsymbol{\alpha}, \omega_0, \lambda_0)$ with $\mathbf{\Sigma} = \mathbf{\Gamma \Phi \Gamma}'$, $\mathbf{S} \backsim \text{MSGHD}(\boldsymbol{\mu}, \mathbf{\Gamma}, \mathbf{\Phi}, \boldsymbol{\alpha}, \boldsymbol{\omega}, \boldsymbol{\lambda})$, and U is an indicator variable such that

$$U = \begin{cases} 1 & \text{if } R \text{ follows a generalized hyperbolic distribution, and} \\ 0 & \text{if } R \text{ follows an MSGHD.} \end{cases}$$

It follows that $\mathbf{X} = \mathbf{\Gamma}\boldsymbol{\mu} + W\mathbf{\Gamma}\boldsymbol{\alpha} + \sqrt{W}\mathbf{\Gamma V}$, where $\mathbf{\Gamma V} \backsim \mathrm{N}_p\left(\mathbf{0}, \mathbf{\Gamma \Phi \Gamma}'\right)$, $\mathbf{S} = \mathbf{\Gamma}\boldsymbol{\mu} + \mathbf{\Gamma}\boldsymbol{\alpha}\mathbf{\Delta_w} + \mathbf{\Gamma A}$, where $\mathbf{\Gamma A} \backsim \mathrm{N}_p\left(\mathbf{0}, \mathbf{\Gamma \Delta_w \Phi \Gamma}'\right)$, and the density of $\mathbf{R}$ can be written

$$\begin{aligned} f_{\text{CGHD}}(\mathbf{r} \mid \boldsymbol{\mu}, \mathbf{\Gamma}, \mathbf{\Phi}, \boldsymbol{\alpha}, \boldsymbol{\omega}, \boldsymbol{\lambda}, \omega_0, \lambda_0, \varpi) = \\ \varpi f_{\text{GHD}} \left(\mathbf{r} \mid \boldsymbol{\mu}, \mathbf{\Gamma \Phi \Gamma}', \boldsymbol{\alpha}, \omega_0, \lambda_0 \right) + (1 - \varpi) f_{\text{MSGHD}} \left(\mathbf{r} \mid \boldsymbol{\mu}, \mathbf{\Gamma}, \mathbf{\Phi}, \boldsymbol{\alpha}, \boldsymbol{\omega}, \boldsymbol{\lambda} \right), \end{aligned} \tag{7.15}$$

where $f_{\text{GHD}}(\cdot)$ is the density of a generalized hyperbolic random variable, $f_{\text{MSGHD}}(\cdot)$ is the density of an MSGHD random variable, and $\varpi \in (0, 1)$ is a mixing proportion. Note that the random vector $\mathbf{R}$ would be distributed generalized hyperbolic if $\varpi = 1$ and would be distributed MSGHD if $\varpi = 0$; however, for identifiability reasons, it is necessary to restrict $\varpi \in (0, 1)$. The parameters $\boldsymbol{\mu}$, $\boldsymbol{\alpha}$, $\mathbf{\Gamma}$, and $\mathbf{\Phi}$ are the same for both densities, the parameters ω_0 and λ_0 are univariate values peculiar to the generalized hyperbolic distribution, and the p-dimensional parameters $\boldsymbol{\omega}$ and $\boldsymbol{\lambda}$ are peculiar to the MSGHD.

Write $\mathbf{R} \curvearrowright \text{CGHD}(\boldsymbol{\mu}, \boldsymbol{\Gamma}, \boldsymbol{\Phi}, \boldsymbol{\alpha}, \boldsymbol{\omega}, \boldsymbol{\lambda}, \omega_0, \lambda_0, \varpi)$ to indicate that the random vector $\mathbf{R}$ follows a CGHD with density in (7.15).

Tortora et al. (2016) use a mixture of CGHDs (MCGHDs) for model-based clustering and classification. The density of an MCGHD is given by

$$f(\mathbf{x} \mid \boldsymbol{\vartheta}) = \sum_{g=1}^{G} \pi_g f_{\text{CGHD}}\left(\mathbf{x} \mid \boldsymbol{\mu}_g, \boldsymbol{\Gamma}_g, \boldsymbol{\Phi}_g, \boldsymbol{\alpha}_g, \boldsymbol{\omega}_g, \boldsymbol{\lambda}_g, \omega_{0g}, \lambda_{0g}, \varpi_g\right).$$

Tortora et al. (2016) use a GEM algorithm for parameter estimation for the MCGHD model. There are four sources of missing data: the latent w_{0ig}, the multi-dimensional weights $\boldsymbol{\Delta}_{\mathbf{w}ig} = \text{diag}(w_{1ig}, \ldots, w_{pig})$, the component membership labels z_{ig}, and inner component labels u_{ig}, for $i = 1, \ldots, n$ and $g = 1, \ldots, G$. As usual, $z_{ig} = 1$ if observation i belongs to component g and $z_{ig} = 0$ otherwise. Similarly, $u_{ig} = 1$ if observation i, in component g, is distributed generalized hyperbolic and $u_{ig} = 0$ if observation i, in component g, is distributed MSGHD. It follows that the (conditional) complete-data log-likelihood for the MCGHD is given by

$$
\begin{aligned}
l_{\text{c}} = \sum_{i=1}^{n} \sum_{g=1}^{G} \Big\{ & z_{ig} \log \pi_g + z_{ig} u_{ig} \log \varpi_g + z_{ig}(1 - u_{ig}) \log(1 - \varpi_g) \\
& + z_{ig} u_{ig} \log h\left(w_{0ig} \mid \omega_{0g}, 1, \lambda_{0g}\right) + z_{ig}(1 - u_{ig}) \sum_{j=1}^{p} \log h\left(w_{jig} \mid \omega_{jg}, 1, \lambda_{jg}\right) \\
& + z_{ig} u_{ig} \log \phi_p\left(\boldsymbol{\Gamma}_g' \mathbf{x}_i \mid \boldsymbol{\mu}_g + w_{0ig} \boldsymbol{\alpha}_g, w_{0ig} \boldsymbol{\Phi}\right) \\
& + z_{ig}(1 - u_{ig}) \sum_{j=1}^{p} \log \phi_1\left([\boldsymbol{\Gamma}_g' \mathbf{x}_i]_j \mid \mu_{jg} + w_{jig} \alpha_{jg}, \omega_{jg} \phi_{jg}\right) \Big\}.
\end{aligned}
$$

Further details on parameter estimation are given by Tortora et al. (2016).

7.6 Cluster Convexity

One theme running through this and the previous chapter is that of additional flexibility. By and large, this additional flexibility has foreseeable and quite obvious consequences. For example, ideas about asymmetric and/or heavier-tailed clusters are quite straightforward and there is relatively limited scope for unintended consequences. The flexibility engendered by the MSGHD dissertation over the generalized hyperbolic distribution, however, deserves some further discussion because, similar to the multiple scaled multivariate-t distribution of Forbes and Wraith (2014), the MSGHD can have contours that are not convex. Accordingly, the MMSGHD can have components that are non-convex, leading to non-convex clusters. The bankruptcy data present a good example of why non-convex clusters can work well in practice (Figure 7.1), and these data will be revisited in Section 7.7.1.

Consider the data in Figure 7.2, which are generated from a mixture of two overlapping t-distributions. When the MMSGHD is fitted to these data for $G = 1, \ldots, 5$ components, the BIC selects a $G = 1$ component model (Figure 7.2) with contours that are clearly non-convex. Looking at the $G = 2$ component MMSGHD solution (Figure 7.2) confirms that the problem is not just one of model selection; the $G = 2$ component MMSGHD solution selects one component that is roughy elliptical and another that seems non-convex.

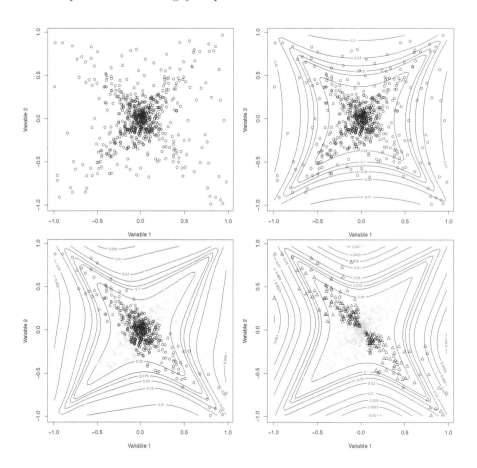

Figure 7.2
Scatter plot of data generated from a two-component t-mixture (top left) along with contours from fitted $G = 1$ component MMSGHD (top right), $G = 2$ component MMSGHD (bottom left), and $G = 2$ component McMS-GHD (bottom right) models, where plotting symbols and shading represent predicted classifications.

Because the MSGHD is susceptible to non-convexity, it is useful to have an analogue that is guaranteed to have convex components. To this end, the constraint $\lambda_j > 1$, $j = 1, \ldots, p$, ensures that the contours are convex and the resulting distribution is called the convex MSGHD (cMSGHD; Tortora et al., 2016). Fitting a mixture of cMSGHDs (McMSGHD) to these data results in a $G = 2$ component model with predicted classifications that are similar to the labels from the generating model (Figure 7.2).

Another example in a similar vein is given in Figure 7.3, where data are generated from a $G = 3$ component mixture of multivariate t-distributions. The selected MMSGHD has $G = 2$ components, including one clearly non-convex cluster, while the McMSGHD gives sensible clustering results.

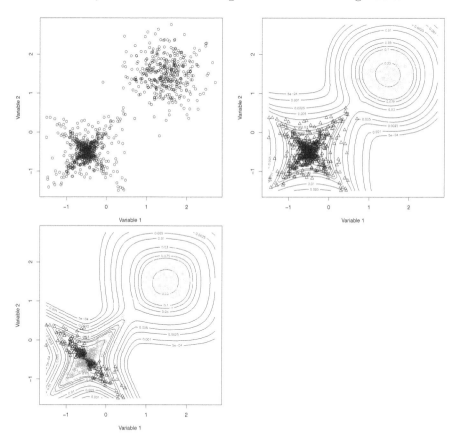

Figure 7.3

Scatter plots for model-based clustering results on data simulated from a three-component t-mixture (top left), with contours from the selected MMS-GHD (top right) and McMSGHD (bottom left) models, respectively, where plotting symbols and shading represent predicted classifications.

The intention behind the introduction of the McMSGHD is not that it should supplant the MMSGHD, but rather that it provides a convenient check on the number of components. In a higher-dimensional application, where visualization is difficult or impossible, situations where the selected MMSGHD has fewer components than the selected McMSGHD will deserve special attention. Of course, this is not to say that the selected McMSGHD will always have more components in situations where the MMSGHD has too few, but rather that it will help to avoid the sort of undesirable situations depicted in Figures 7.2 and 7.3.

7.7   Illustrations

7.7.1   Bankruptcy Data

Return to the `bankruptcy` data from the `MixGHD` package, which contain the ratio of retained earnings to total assets as well as the ratio of earnings before interest and taxes to total assets for 66 American firms. The MCGHD is fitted to these data for $G = 2$, using k-medoids starting values. The resulting model gives predicted classifications that are close to the true classes (Figure 7.4).

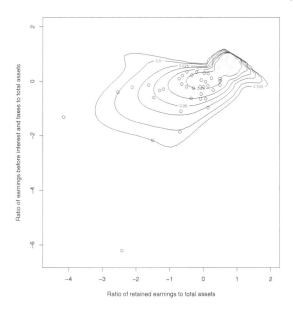

Figure 7.4
Scatter plot for the (scaled) bankruptcy data, with contours from a fitted MCGHD, where shading and plotting symbols represent predicted classifications.

The MGHD, MMSGHD, and McMSGHD models were also fitted to these data, for $G = 2$, using k-medoids starts. For the MMSGHD and McMSGHD models, the associated MAP classifications do not correspond to bankruptcy status (Figure 7.5). Rather, one component consists largely of the bankrupts with both ratios in similarly poor health and the other component takes in the remaining bankrupts as well as almost all of the financially sound firms — although it does not correspond to status, it could be argued that this is nonetheless an interesting clustering solution. For the MGHD, the associated MAP classifications are different from those for the MMSGHD and McMSGHD models; however, they do not correspond to bankruptcy status (Figure 7.5). The results for the bankruptcy data illustrate that the MCGHD approach can give very good classification performance in situations where neither the MGHD nor the MMSGHD produce similar results.

Finally, it is interesting to observe the fitted $G = 2$ component MMSGHD model for the bankruptcy data with all points treated as labelled (Figure 7.6). This illustrates the potential of multiple scaled distributions for density estimation and, by extension, discriminant analysis.

7.7.2   Other Clustering Examples

In addition to the bankruptcy data, two other datasets are considered for clustering. The `banknote` data from `mclust` contain five measurements on 100 genuine and 100 counterfeit Swiss 1,000-franc banknotes. The Australian Institute of Sport (AIS) data contain 11 biometric measurements on 102 male and 100 female athletes (Table 7.1), and are available in the `alr3` package for R.

Table 7.1
Eleven biometric measurements from the AIS data.

Height (cm)	White cell count	Body mass index
Weight (kg)	Hematocrit	Sum of skin folds
Lean body mass	Hemoglobin	Percent body fat
Red cell count	Plasma ferritin concentration	

For each dataset, four clustering approaches — MCGHD, MGHD, MMS-GHD, and McMSGHD — are fitted with the number of components set equal to the true number of classes. The results are summarized in Table 7.2, along with those for the bankruptcy data. Notably, the MCGHD performs very well for all datasets and is the only approach to perform well on the bankruptcy data. The other approaches give very good classification performance on the banknote and AIS datasets but do not perform well on the bankruptcy data. Interestingly, the McMSGHD approach slightly outperforms MMSGHD for the AIS data. Also, the MGHD approach performs as well as the MMSGHD

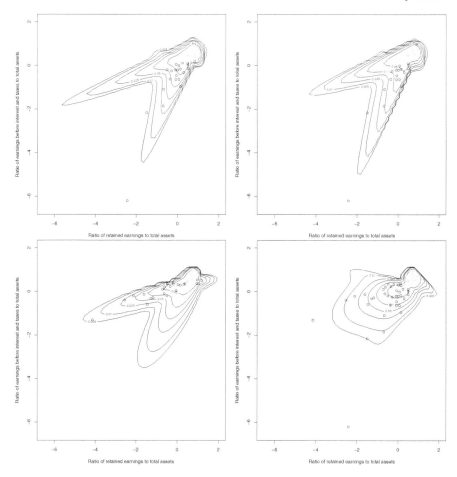

Figure 7.5
Scatter plots of the (scaled) bankruptcy data, with contours from the fitted two-component MMSGHD (top left), McMSGHD (top right), MGHD (bottom left), and MCGHD (bottom right), respectively, where plotting symbols and shading represent predicted classifications.

for the banknote and AIS data. This highlights the fact that a mixture of multiple scaled distributions may well not outperform its single scaled analogue. In fact, as will be shown in the next section (Table 7.3), it may even give inferior classification performance; this underlines the need for an approach with both the MMSGHD and MGHD models as special or limiting cases.

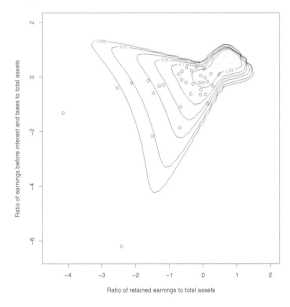

Figure 7.6
Scatter plot for the (scaled) bankruptcy data, with contours from a fitted MMSGHD, where all points were treated as labelled, and plotting symbol and shading represent status, i.e., filed for bankruptcy or financially sound.

Table 7.2
ARI values for model-based clustering analyses using two-component MCGHD, MGHD, MMSGHD, and McMSGHD models, respectively.

	MCGHD	MGHD	MMSGHD	McMSGHD
Banknote	0.980	0.980	0.980	0.980
Bankruptcy	0.824	0.061	0.255	0.170
AIS	0.903	0.884	0.884	0.903

7.7.3 Classification and Discriminant Analysis Examples

Three datasets are considered: the bankruptcy data, the diabetes data from `mclust`, and the Italian olive oil data from `pgmm`. In each case, a subset of around 25% of the observations is treated as unlabelled and all four approaches — MCGHD, MGHD, MMSGHD, and McMSGHD — are fitted. Note that random starts are used for the unlabelled observations in the model-based classification analyses, and only one component per class is used for discriminant analysis. The results are summarized, via ARI values, in Tables 7.3 and 7.4. Because the bankruptcy data contain only 66 firms in total, the unlabelled subset contains just 16 firms and so the ARI values need to be interpreted carefully, i.e., the classification performance is very good despite seemingly poor ARI values. For example, the MCGHD misclassifies just two out of 16 unlabelled firms (Table 7.5) corresponding to an ARI value of 0.53. All four approaches perform very well for the olive oil data and diabetes data. For discriminant analysis, the MCGHD, MMSGHD, and MGHD approaches perform very well on all three datasets. The McMSGHD approach gives somewhat inferior performance on the diabetes data but performs very well on the bankruptcy and olive oil datasets.

Table 7.3
ARI values, for unlabelled observations, associated with model-based classification analysis on three real datasets using MCGHD, MGHD, MMSGHD, and McMSGHD distributions, respectively.

	MCGHD	**MGHD**	**MMSGHD**	**McMSGHD**
Bankruptcy	0.533	0.533	0.533	0.533
Diabetes	0.954	0.915	0.784	0.834
Olive oil	0.958	0.961	0.940	0.923

Table 7.4
ARI values, for unlabelled observations, associated with discriminant analysis on three real datasets using MCGHD, MGHD, MMSGHD, and McMSGHD distributions, respectively.

	MCGHD	**MGHD**	**MMSGHD**	**McMSGHD**
Bankruptcy	1.000	1.000	1.000	1.000
Diabetes	0.825	0.789	0.825	0.662
Olive oil	0.888	0.876	0.888	0.890

Table 7.5
Cross-tabulation of the MAP classifications (A, B) associated with the fitted MCGHD model against true classes for the `bankrutcy` data, where 16 firms are treated as unlabelled.

	A	**B**
Filed for bankruptcy	8	0
Financially sound	2	6

7.8   Comments

Mixtures of multiple scaled distributions are a recent addition (at the time of writing) to the literature on mixture model-based approaches to clustering, classification, and discriminant analysis. The results of the cluster analyses carried out in this chapter (cf. Table 7.2) highlight some important points. For one, the analysis of the bankruptcy data highlights that the MCGHD can give very good clustering performance in scenarios where the MGHD, MMSGHD, and McMSGHD distributions all perform poorly. This is particularly notable in the case of the MGHD and MMSGHD because the MCGHD density arises as a convex combination of their respective densities, cf. (7.15). It is also noteworthy that the MCGHD is the only approach applied to these three datasets that does not give poor clustering performance in any case (ARI ≥ 0.824). Perhaps the most significant point to emerge from these cluster analyses, however, arises from comparison of the MGHD and the MMSGHD results: both do well for the banknote and AIS datasets, and both are poor for the bankruptcy data. This underlines that, while multiple scaled distributions can bring great flexibility, they should not be expected to necessarily outperform their single scaled analogues. This point is further reinforced by the model-based classification results, where the MGHD outperforms the MMSGHD for two of the three datasets (Table 7.3).

The model-based classification and discriminant analyses results presented in Tables 7.3 and 7.4 raise other interesting points. Considering how poorly Gaussian model-based discriminant analysis performed on the olive oil data (cf. Section 2.6.4), it is notable that all four approaches gave excellent discriminant analysis results on these data (ARI ≥ 0.876). This illustrates that it can be much better to use one suitable density per class rather than fitting multiple Gaussian densities for some classes.

Another important point to emerge from this work is the potential importance of cluster convexity. This point is sometimes overlooked and the consequences of doing so can be most unfortunate (e.g., Figures 7.2 and 7.3). This said, it can sometimes be desirable to have one or more components that are not convex; the bankruptcy data are a good example of a situation where (at least) one non-convex component is needed.

Finally, there are unanswered questions about choosing starting values and model selection criteria for the mixtures of multiple scaled distributions discussed in this chapter. In the real data illustrations in Section 7.7, k-medoids starting values were used for clustering with G fixed to equal the number of classes. However, much more work is required to determine an optimum approach for selection of starting values and model selection; see Tortora et al. (2016) for a comparison of some model selection criteria.

8

Methods for Longitudinal Data

8.1 Modified Cholesky Decomposition

The Cholesky decomposition (Benoît, 1924) is a method for decomposing a matrix into the product of a lower triangular matrix and its transpose. As an aside, note that Cholesky died in World War I and his method was published posthumously by a colleague from the French military (Benoît). Let $\mathbf{A}$ be a real, positive definite matrix; then the Cholesky decomposition of $\mathbf{A}$ is

$$\mathbf{A} = \mathbf{L}\mathbf{L}', \tag{8.1}$$

where $\mathbf{L}$ is a unique lower triangular matrix. The decomposition in (8.1) is often used in numerical analysis applications to simplify the solution to a system of linear equations.

Pourahmadi (1999, 2000) applies a modified Cholesky decomposition to the covariance matrix $\mathbf{\Sigma}$ of a random variable, i.e.,

$$\mathbf{T}\mathbf{\Sigma}\mathbf{T}' = \mathbf{D}, \tag{8.2}$$

where $\mathbf{T}$ is a unique unit lower triangular matrix and $\mathbf{D}$ is a unique diagonal matrix with strictly positive diagonal entries. Note that a unit lower triangular matrix is a lower triangular matrix with diagonal elements 1. Alternatively, the relationship in (8.2) can be written

$$\mathbf{\Sigma}^{-1} = \mathbf{T}'\mathbf{D}^{-1}\mathbf{T}. \tag{8.3}$$

The values of $\mathbf{T}$ and $\mathbf{D}$ can be interpreted as generalized autoregressive parameters and innovation variances, respectively (Pourahmadi, 1999), so that the linear least-squares predictor of X_t, based on $X_{t-1}, \ldots, X_1$, is given by

$$\hat{X}_t = \mu_t + \sum_{s=1}^{t-1} (-\varphi_{ts})(X_s - \mu_s) + \sqrt{d_t}\varepsilon_t, \tag{8.4}$$

where φ_{ts} is the sub-diagonal element of $\mathbf{T}$ in position (t, s), d_t is the tth diagonal element of $\mathbf{D}$, and $\varepsilon_t \sim \mathrm{N}(0, 1)$.

Pan and MacKenzie (2003) use the modified Cholesky decomposition for joint modelling of the mean and covariance in longitudinal studies. Pourahmadi et al. (2007) develop an approach for simultaneously modelling several covariance matrices via this decomposition, thereby giving an alternative to common principal components analysis (Flury, 1988) for longitudinal data.

8.2   Gaussian Mixture Modelling of Longitudinal Data

8.2.1   The Model

McNicholas and Murphy (2010a) use a Gaussian mixture model with a modified Cholesky-decomposed covariance structure for each component to model longitudinal data. They consider model-based clustering applications, but model-based classification and model-based discriminant analysis proceed in an analogous fashion. Consider a Gaussian mixture model with the decomposition in (8.3) used for each component precision matrix Σ_g^{-1} so that the gth component density, for a p-dimensional random variable $\mathbf{X}$, is given by

$$\phi(\mathbf{x} \mid \boldsymbol{\mu}_g, (\mathbf{T}_g' \mathbf{D}_g^{-1} \mathbf{T}_g)^{-1}) =$$
$$\frac{1}{\sqrt{(2\pi)^p |\mathbf{D}_g|}} \exp\left\{ -\frac{1}{2} (\mathbf{x} - \boldsymbol{\mu}_g)' \mathbf{T}_g' \mathbf{D}_g^{-1} \mathbf{T}_g (\mathbf{x} - \boldsymbol{\mu}_g) \right\},$$

where $\mathbf{T}_g$ and $\mathbf{D}_g$ are the $p \times p$ unit lower triangular matrix and the $p \times p$ diagonal matrix, respectively, that follow from the modified Cholesky decomposition of Σ_g.

A family of eight Gaussian mixture models arises from the option to constrain $\mathbf{T}_g$ and/or $\mathbf{D}_g$ to be equal across components together with the option to impose the isotropic constraint $\mathbf{D}_g = \delta_g \mathbf{I}_p$. This family will be called the Cholesky-decomposed Gaussian mixture model (CDGMM) family herein. Each member of the CDGMM family (Table 8.1) has a natural interpretation for longitudinal data. Constraining $\mathbf{T}_g = \mathbf{T}$ suggests that the autoregressive relationship between time points, cf. (8.4), is the same across components. The constraint $\mathbf{D}_g = \mathbf{D}$ suggests that the variability at each time point is the same for each component, and the isotropic constraint $\mathbf{D}_g = \delta_g \mathbf{I}_p$ suggests that the variability is the same at each time point in component g.

Table 8.1
The nomenclature, covariance structure, and number of free covariance parameters for each member of the CDGMM family.

Model	$\mathbf{T}_g$	$\mathbf{D}_g$	$\mathbf{D}_g$	Free Cov. Parameters
EEA	Equal	Equal	Anisotropic	$p(p-1)/2 + p$
VVA	Variable	Variable	Anisotropic	$G[p(p-1)/2] + Gp$
VEA	Variable	Equal	Anisotropic	$G[p(p-1)/2] + p$
EVA	Equal	Variable	Anisotropic	$p(p-1)/2 + Gp$
VVI	Variable	Variable	Isotropic	$G[p(p-1)/2] + G$
VEI	Variable	Equal	Isotropic	$G[p(p-1)/2] + 1$
EVI	Equal	Variable	Isotropic	$p(p-1)/2 + G$
EEI	Equal	Equal	Isotropic	$p(p-1)/2 + 1$

From a clustering or classification point of view, two of the models in Table 8.1 have equivalent GPCM models (Table 2.1). Specifically, the EEA model from Table 8.1 is equivalent to the GPCM model EEE, i.e., unconstrained, common component covariance matrices, and the VVA model from Table 8.1 is equivalent to the GPCM model VVV, i.e., unconstrained, differing component covariance matrices. Of course, even though this equivalence exists, the GPCM models do not explicitly account for the longitudinal correlation structure and the models in Table 8.1 are more natural for longitudinal data.

8.2.2 Model Fitting

8.2.2.1 VEA Model

Each of the models in Table 8.1 can be fitted using an EM algorithm. Assuming a clustering paradigm, the complete-data comprise the observed $\mathbf{x}_1, \ldots, \mathbf{x}_n$ together with the unknown labels $\mathbf{z}_1, \ldots, \mathbf{z}_n$, where $\mathbf{z}_i = (z_{i1}, \ldots, z_{iG})$ and $z_{ig} = 1$ if observation i belongs to component g and $z_{ig} = 0$ otherwise. First, consider the VEA model, so that $\mathbf{D}_g = \mathbf{D}$. The complete-data log-likelihood is

$$l_c(\boldsymbol{\vartheta}) = \sum_{i=1}^{n} \sum_{g=1}^{G} z_{ig} \left[\log \pi_g + \log \phi(\mathbf{x}_i \mid \boldsymbol{\mu}_g, (\mathbf{T}_g' \mathbf{D}^{-1} \mathbf{T}_g)^{-1}) \right],$$

where $\boldsymbol{\vartheta}$ denotes the model parameters. Replacing the z_{ig} by their expected values, i.e.,

$$\hat{z}_{ig} = \frac{\hat{\pi}_g \phi(\mathbf{x}_i \mid \hat{\boldsymbol{\mu}}_g, (\hat{\mathbf{T}}_g' \hat{\mathbf{D}}^{-1} \hat{\mathbf{T}}_g)^{-1})}{\sum_{h=1}^{G} \hat{\pi}_h \phi(\mathbf{x}_i \mid \hat{\boldsymbol{\mu}}_h, (\hat{\mathbf{T}}_h' \hat{\mathbf{D}}^{-1} \hat{\mathbf{T}}_h)^{-1})},$$

gives the expected value of the complete-data log-likelihood, i.e.,

$$
\begin{aligned}
Q = \sum_{g=1}^{G} n_g \log \pi_g - \frac{np}{2} \log 2\pi - \sum_{g=1}^{G} \frac{n_g}{2} \log \left| (\mathbf{T}_g' \mathbf{D}^{-1} \mathbf{T}_g)^{-1} \right| \\
- \sum_{g=1}^{G} \frac{n_g}{2} \operatorname{tr} \left\{ \mathbf{S}_g \mathbf{T}_g' \mathbf{D}^{-1} \mathbf{T}_g \right\},
\end{aligned}
\tag{8.5}
$$

where $n_g = \sum_{i=1}^{n} \hat{z}_{ig}$ and

$$\mathbf{S}_g = \frac{1}{n_g} \sum_{i=1}^{n} \hat{z}_{ig} (\mathbf{x}_i - \boldsymbol{\mu}_g)(\mathbf{x}_i - \boldsymbol{\mu}_g)'.$$

Note that, as usual, we are conditioning on the current parameter estimates. Now, maximizing (8.5) with respect to π_g and $\boldsymbol{\mu}_g$ gives

$$\hat{\pi}_g = \frac{n_g}{n} \qquad \text{and} \qquad \hat{\boldsymbol{\mu}}_g = \frac{\sum_{i=1}^{n} \hat{z}_{ig} \mathbf{x}_i}{n_g},$$

respectively.

Using some basic linear algebra results (cf. Appendix A.1), the term

$$\sum_{g=1}^{G} \frac{n_g}{2} \log \left|(\mathbf{T}_g' \mathbf{D}^{-1} \mathbf{T}_g)^{-1}\right|$$

in (8.5) can be simplified via:

$$\log \left|(\mathbf{T}_g' \mathbf{D}^{-1} \mathbf{T}_g)^{-1}\right| = \log \left|\mathbf{T}_g' \mathbf{D}^{-1} \mathbf{T}_g\right|^{-1} = \log \left(|\mathbf{T}_g|^2 \, |\mathbf{D}^{-1}|\right)^{-1} = \log |\mathbf{D}|.$$

Therefore,

$$\sum_{g=1}^{G} \frac{n_g}{2} \log \left|(\mathbf{T}_g' \mathbf{D}^{-1} \mathbf{T}_g)^{-1}\right| = \sum_{g=1}^{G} \frac{n_g}{2} \log |\mathbf{D}| = \frac{n}{2} \log |\mathbf{D}|,$$

and, noting that

$$\mathrm{tr}\left\{\mathbf{S}_g \mathbf{T}_g' \mathbf{D}^{-1} \mathbf{T}_g\right\} = \mathrm{tr}\left\{\mathbf{T}_g \mathbf{S}_g \mathbf{T}_g' \mathbf{D}^{-1}\right\},$$

(8.5) can be written

$$Q = C - \frac{n}{2} \log |\mathbf{D}| - \sum_{g=1}^{G} \frac{n_g}{2} \mathrm{tr}\left\{\mathbf{T}_g \mathbf{S}_g \mathbf{T}_g' \mathbf{D}^{-1}\right\}, \tag{8.6}$$

where C is constant with respect to $\mathbf{T}_g$ and $\mathbf{D}$. Differentiating (8.6) with respect to $\mathbf{T}_g$ and $\mathbf{D}^{-1}$, respectively, gives the score functions

$$S_1(\mathbf{T}_g, \mathbf{D}) = \frac{\partial Q}{\partial \mathbf{T}_g} = -\frac{n_g}{2} \left[(\mathbf{D}^{-1})' \mathbf{T}_g \mathbf{S}_g' + \mathbf{D}^{-1} \mathbf{T}_g \mathbf{S}_g\right] = -n_g \mathbf{D}^{-1} \mathbf{T}_g \mathbf{S}_g,$$

$$S_2(\mathbf{T}_g, \mathbf{D}) = \frac{\partial Q}{\partial \mathbf{D}^{-1}} = \frac{n}{2} \mathbf{D} - \sum_{g=1}^{G} \frac{n_g}{2} (\mathbf{T}_g \mathbf{S}_g \mathbf{T}_g')' = \frac{n}{2} \mathbf{D} - \sum_{g=1}^{G} \frac{n_g}{2} \mathbf{T}_g \mathbf{S}_g \mathbf{T}_g'.$$

Let $\varphi_{ij}^{(g)}$ denote those elements of $\mathbf{T}_g$ that are to be estimated, so that

$$\mathbf{T}_g = \begin{pmatrix} 1 & 0 & 0 & 0 & \cdots & 0 \\ \varphi_{21}^{(g)} & 1 & 0 & 0 & \cdots & 0 \\ \varphi_{31}^{(g)} & \varphi_{32}^{(g)} & 1 & 0 & \cdots & 0 \\ \vdots & \vdots & & \ddots & & \vdots \\ \varphi_{p-1,1}^{(g)} & \varphi_{p-1,2}^{(g)} & \cdots & \varphi_{p-1,p-2}^{(g)} & 1 & 0 \\ \varphi_{p1}^{(g)} & \varphi_{p3}^{(g)} & \cdots & \varphi_{p,p-2}^{(g)} & \varphi_{p,p-1}^{(g)} & 1 \end{pmatrix}. \tag{8.7}$$

Define $\Phi_g = \{\varphi_{ij}^{(g)}\}$ for $i > j$ with $i, j \in \{1, 2, \ldots, p\}$, and let $\mathrm{LT}\{\cdot\}$ denote the lower triangular part of a matrix.

Now, solving $S_1(\hat{\mathbf{T}}_g, \mathbf{D}) \equiv \text{LT}\{S_1(\hat{\Phi}_g, \mathbf{D})\} = \mathbf{0}$ for $\hat{\Phi}_g$ leads to a total of $p - 1$ systems of linear equations. The 1×1 system is straightforward:

$$\frac{s_{11}^{(g)} \hat{\varphi}_{21}^{(g)}}{d_2} + \frac{s_{21}^{(g)}}{d_2} = 0,$$

and so

$$\hat{\varphi}_{21}^{(g)} = -\frac{s_{21}^{(g)}}{s_{11}^{(g)}},$$

where $s_{ij}^{(g)}$ denotes the element in row i and column j of $\mathbf{S}_g$ and d_i denotes the ith diagonal element of $\mathbf{D}$. The 2×2 system is

$$\frac{s_{11}^{(g)} \hat{\varphi}_{31}^{(g)}}{d_3} + \frac{s_{21}^{(g)} \hat{\varphi}_{32}^{(g)}}{d_3} + \frac{s_{31}^{(g)}}{d_3} = 0,$$

$$\frac{s_{12}^{(g)} \hat{\varphi}_{31}^{(g)}}{d_3} + \frac{s_{22}^{(g)} \hat{\varphi}_{32}^{(g)}}{d_3} + \frac{s_{32}^{(g)}}{d_3} = 0,$$

which can be expressed in the form

$$\begin{pmatrix} s_{11}^{(g)} & s_{21}^{(g)} \\ s_{12}^{(g)} & s_{22}^{(g)} \end{pmatrix} \begin{pmatrix} \hat{\varphi}_{31}^{(g)} \\ \hat{\varphi}_{32}^{(g)} \end{pmatrix} = -\begin{pmatrix} s_{31}^{(g)} \\ s_{32}^{(g)} \end{pmatrix}, \tag{8.8}$$

and so

$$\begin{pmatrix} \hat{\varphi}_{31}^{(g)} \\ \hat{\varphi}_{32}^{(g)} \end{pmatrix} = -\begin{pmatrix} s_{11}^{(g)} & s_{21}^{(g)} \\ s_{12}^{(g)} & s_{22}^{(g)} \end{pmatrix}^{-1} \begin{pmatrix} s_{31}^{(g)} \\ s_{32}^{(g)} \end{pmatrix}. \tag{8.9}$$

Proceeding in this way, it follows that the solution to the $(r - 1) \times (r - 1)$ system of equations is

$$\begin{pmatrix} \hat{\varphi}_{r1}^{(g)} \\ \hat{\varphi}_{r2}^{(g)} \\ \vdots \\ \hat{\varphi}_{r,r-1}^{(g)} \end{pmatrix} = -\begin{pmatrix} s_{11}^{(g)} & s_{21}^{(g)} & \cdots & s_{r-1,1}^{(g)} \\ s_{12}^{(g)} & s_{22}^{(g)} & \cdots & s_{r-1,2}^{(g)} \\ \vdots & \vdots & \ddots & \vdots \\ s_{1,r-1}^{(g)} & s_{2,r-2}^{(g)} & \cdots & s_{r-1,r-1}^{(g)} \end{pmatrix}^{-1} \begin{pmatrix} s_{r1}^{(g)} \\ s_{r2}^{(g)} \\ \vdots \\ s_{r,r-1}^{(g)} \end{pmatrix}, \tag{8.10}$$

for $r = 2, \ldots, p$. Solving $\text{diag}\{S_2(\hat{\mathbf{T}}_g, \hat{\mathbf{D}})\} = \mathbf{0}$ gives

$$\hat{\mathbf{D}} = \sum_{g=1}^{G} \hat{\pi}_g \, \text{diag}\{\hat{\mathbf{T}}_g \mathbf{S}_g \hat{\mathbf{T}}_g'\}. \tag{8.11}$$

Parameter estimation for most of the other models is similar to that for the VEA model, with the two exceptions being the EVA and EVI models. Details for the EVA model are given by McNicholas and Murphy (2010a) and details for the EVI model are provided in Section 8.2.2.2.

8.2.2.2   EVI Model

Again, assume a clustering paradigm, so that the complete-data comprise the observed $\mathbf{x}_1, \ldots, \mathbf{x}_n$ together with the unknown group membership labels $\mathbf{z}_1, \ldots, \mathbf{z}_n$. Now consider the EVI model, so that $\mathbf{T}_g = \mathbf{T}$ and $\mathbf{D}_g = \delta_g \mathbf{I}_p$. The complete-data log-likelihood is given by

$$l_c(\boldsymbol{\vartheta}) = \sum_{i=1}^{n} \sum_{g=1}^{G} z_{ig} \left[\log \pi_g + \log \phi(\mathbf{x}_i \mid \boldsymbol{\mu}_g, (\delta_g^{-1} \mathbf{T}' \mathbf{T})^{-1}) \right],$$

where $\boldsymbol{\vartheta}$ denotes the model parameters. Replacing the z_{ig} by their expected values, i.e.,

$$\hat{z}_{ig} = \frac{\hat{\pi}_g \phi(\mathbf{x}_i \mid \hat{\boldsymbol{\mu}}_g, (\hat{\delta}_g^{-1} \hat{\mathbf{T}}' \hat{\mathbf{T}})^{-1})}{\sum_{h=1}^{G} \hat{\pi}_h \phi(\mathbf{x}_i \mid \hat{\boldsymbol{\mu}}_h, (\hat{\delta}_h^{-1} \hat{\mathbf{T}}' \hat{\mathbf{T}})^{-1})},$$

gives the expected value of the complete-data log-likelihood, conditional on the current parameter estimates, i.e.,

$$\begin{aligned}
Q &= C + \sum_{g=1}^{G} n_g \log \pi_g - \sum_{g=1}^{G} \frac{n_g}{2} \log |\delta_g \mathbf{I}_p| - \sum_{g=1}^{G} \frac{n_g}{2} \operatorname{tr} \left\{ \mathbf{T} \mathbf{S}_g \mathbf{T}' \delta_g^{-1} \right\} \\
&= C + \sum_{g=1}^{G} n_g \log \pi_g + \sum_{g=1}^{G} \frac{n_g p}{2} \log \delta_g^{-1} - \sum_{g=1}^{G} \frac{n_g \delta_g^{-1}}{2} \operatorname{tr} \left\{ \mathbf{T} \mathbf{S}_g \mathbf{T}' \right\},
\end{aligned} \tag{8.12}$$

where C is constant with respect to the model parameters, and n_g and $\mathbf{S}_g$ are as defined before (Section 8.2.2.1).

Maximizing (8.12) with respect to π_g and $\boldsymbol{\mu}_g$ gives

$$\hat{\pi}_g = \frac{n_g}{n} \qquad \text{and} \qquad \hat{\boldsymbol{\mu}}_g = \frac{\sum_{i=1}^{n} \hat{z}_{ig} \mathbf{x}_i}{n_g},$$

respectively. Differentiating (8.12) with respect to $\mathbf{T}$ and δ_g^{-1}, respectively, gives the following score functions:

$$S_1(\mathbf{T}, \delta_g) = \frac{\partial Q}{\partial \mathbf{T}} = -\sum_{g=1}^{G} \frac{n_g \delta_g^{-1}}{2} \left(\mathbf{T} \mathbf{S}_g' + \mathbf{T} \mathbf{S}_g \right) = -\sum_{g=1}^{G} n_g \delta_g^{-1} \mathbf{T} \mathbf{S}_g,$$

$$S_2(\mathbf{T}, \delta_g) = \frac{\partial Q}{\partial \delta_g^{-1}} = \frac{n_g p}{2} \delta_g - \frac{n_g}{2} \operatorname{tr}\{\mathbf{T} \mathbf{S}_g \mathbf{T}'\} = \frac{n_g}{2} \left(p \delta_g - \operatorname{tr}\{\mathbf{T} \mathbf{S}_g \mathbf{T}'\} \right).$$

Again, let φ_{ij} denote those elements of $\mathbf{T}$ that are to be estimated. Define $\Phi = \{\varphi_{ij}\}$ for $i > j$ with $i, j \in \{1, 2, \ldots, p\}$, and use $\operatorname{LT}\{\cdot\}$, as before, to denote the lower triangular part of a matrix. Now, solving $\operatorname{LT}\{S_1(\hat{\Phi}, \delta_g)\} = \mathbf{0}$ for $\hat{\Phi}$ leads again to a total of $p - 1$ systems of linear equations. The 1×1 system is

$$\sum_{g=1}^{G} \left[\frac{s_{11}^{(g)} \hat{\varphi}_{21}}{\delta_g} + \frac{s_{21}^{(g)}}{\delta_g} \right] = 0,$$

and so

$$\hat{\varphi}_{21} = -\left[\sum_{g=1}^{G} \frac{s_{21}^{(g)}}{\delta_g}\right]\left[\sum_{g=1}^{G} \frac{s_{11}^{(g)}}{\delta_g}\right]^{-1}. \tag{8.13}$$

Using similar notation to McNicholas and Murphy (2010a), write

$$\kappa^{ij} = \sum_{g=1}^{G} \frac{s_{ij}^{(g)}}{\delta_g},$$

so that (8.13) can be written $\hat{\varphi}_{21} = -\kappa^{21}/\kappa^{11}$. The 2×2 system is

$$\kappa^{11}\hat{\varphi}_{31} + \kappa^{21}\hat{\varphi}_{32} + \kappa^{31} = 0,$$
$$\kappa^{12}\hat{\varphi}_{31} + \kappa^{22}\hat{\varphi}_{32} + \kappa^{32} = 0,$$

which can be written

$$\begin{pmatrix} \kappa^{11} & \kappa^{21} \\ \kappa^{12} & \kappa^{22} \end{pmatrix}\begin{pmatrix} \hat{\varphi}_{31} \\ \hat{\varphi}_{32} \end{pmatrix} = -\begin{pmatrix} \kappa^{31} \\ \kappa^{32} \end{pmatrix},$$

and so

$$\begin{pmatrix} \hat{\varphi}_{31} \\ \hat{\varphi}_{32} \end{pmatrix} = -\begin{pmatrix} \kappa^{11} & \kappa^{21} \\ \kappa^{12} & \kappa^{22} \end{pmatrix}^{-1}\begin{pmatrix} \kappa^{31} \\ \kappa^{32} \end{pmatrix}.$$

In general, the solution to the $(r-1) \times (r-1)$ system of equations is

$$\begin{pmatrix} \hat{\varphi}_{r1} \\ \hat{\varphi}_{r2} \\ \vdots \\ \hat{\varphi}_{r,r-1}^{(g)} \end{pmatrix} = -\begin{pmatrix} \kappa^{11} & \kappa^{21} & \cdots & \kappa^{r-1,1} \\ \kappa^{12} & \kappa^{22} & \cdots & \kappa^{r-1,2} \\ \vdots & \vdots & \ddots & \vdots \\ \kappa^{1,r-1} & \kappa^{2,r-2} & \cdots & \kappa^{r-1,r-1} \end{pmatrix}^{-1}\begin{pmatrix} \kappa^{r1} \\ \kappa^{r2} \\ \vdots \\ \kappa^{r,r-1} \end{pmatrix},$$
$$\tag{8.14}$$

for $r = 2, 3, \ldots, p$. Note that, because $\kappa^{ij} = \kappa^{ji}$, the $(r-1) \times (r-1)$ matrix in (8.14) is symmetric. Now, solving $S_2(\hat{\mathbf{T}}, \hat{\delta}_g) = 0$ gives

$$\hat{\delta}_g = \frac{1}{p}\text{tr}\{\hat{\mathbf{T}}\mathbf{S}_g\hat{\mathbf{T}}'\}. \tag{8.15}$$

8.2.3 Constraining Sub-Diagonals of $\mathbf{T}_g$

8.2.3.1 V_dEA Model

McNicholas and Murphy (2010a) also consider cases where elements below a given sub-diagonal of $\mathbf{T}_g$ are set to zero. This constrained correlation structure removes autocorrelation over large time lags, i.e., constraining $\mathbf{T}_g$ to zero below the dth sub-diagonal implies an order d autoregressive structure within the framework of (8.4). The notation V_dEA is used to denote the VEA model where the elements of $\mathbf{T}_g$ are zero below the dth sub-diagonal. Note that

McNicholas and Murphy (2010a) consider $d \geq 1$, pointing out that models with $d = 0$ would be equivalent to diagonal GPCM models.

Parameter estimation when only the first sub-diagonal of $\mathbf{T}_g$ is non-zero, i.e., for $d = 1$, is straightforward. For example, from the work in Section 8.2.2.1, the update for $\mathbf{T}_g$ in the M-step of the V_1EA model is

$$\hat{\varphi}_{r,r-1}^{(g)} = -\frac{s_{r,r-1}^{(g)}}{s_{r-1,r-1}^{(g)}},$$

for $r = 2, \ldots, p$. Next, consider the V_2EA model. Let $SD_r\{\cdot\}$ denote the first r sub-diagonals of a matrix. Solving $SD_2\{S_1(\hat{\mathbf{\Phi}}_g, \mathbf{D})\} = 0$ for $\hat{\mathbf{\Phi}}_g$ leads to a total of $p - 1$ systems of linear equations. One of these systems is 1×1, and gives the familiar solution

$$\hat{\varphi}_{21}^{(g)} = -\frac{s_{21}^{(g)}}{s_{11}^{(g)}}.$$

The other systems are all 2×2, with solutions given by

$$\begin{pmatrix} \hat{\varphi}_{r,r-2}^{(g)} \\ \hat{\varphi}_{r,r-1}^{(g)} \end{pmatrix} = - \begin{pmatrix} s_{r-2,r-2}^{(g)} & s_{r-1,r-2}^{(g)} \\ s_{r-2,r-1}^{(g)} & s_{r-1,r-1}^{(g)} \end{pmatrix}^{-1} \begin{pmatrix} s_{r,r-2}^{(g)} \\ s_{r,r-1}^{(g)} \end{pmatrix},$$

for $r = 3, \ldots, p$, and solving $\text{diag}\{S_2(\hat{\mathbf{T}}_g, \hat{\mathbf{D}})\} = \mathbf{0}$ for $\hat{\mathbf{D}}$ gives the same solution as in (8.11).

The updates for $\mathbf{T}_g$ and $\mathbf{D}$ can be generalized to the V_dEA case. Using the same notation, the $\hat{\mathbf{\Phi}}_g$ are given by

$$\begin{pmatrix} \hat{\varphi}_{r,r-d}^{(g)} \\ \hat{\varphi}_{r,r-(d-1)}^{(g)} \\ \vdots \\ \hat{\varphi}_{r,r-1}^{(g)} \end{pmatrix} =$$

$$- \begin{pmatrix} s_{r-d,r-d}^{(g)} & s_{r-(d-1),r-d}^{(g)} & \cdots & s_{r-1,r-d}^{(g)} \\ s_{r-d,r-(d-1)}^{(g)} & s_{r-(d-1),r-(d-1)}^{(g)} & \cdots & s_{r-1,r-(d-1)}^{(g)} \\ \vdots & \vdots & \ddots & \vdots \\ s_{r-d,r-1}^{(g)} & s_{r-(d-1),r-1}^{(g)} & \cdots & s_{r-1,r-1}^{(g)} \end{pmatrix}^{-1} \begin{pmatrix} s_{r,r-d}^{(g)} \\ s_{r,r-(d-1)}^{(g)} \\ \vdots \\ s_{r,r-1}^{(g)} \end{pmatrix},$$

for $r = 2, \ldots, p$ and the update for $\hat{\mathbf{D}}$ is as in (8.11). Parameter estimation for five of the other models, with the sub-diagonal constraints on the $\mathbf{T}_g$, are similar to those for the V_dEA model shown here. Again, the two exceptions are the E_dVA and E_dVI models. Updates for the E_dVA case are given by McNicholas and Murphy (2010a) and updates for the E_dVI model follow here.

8.2.3.2 E_dVI Model

First consider the E_2VI model and recall work on the EVI model from Section 8.2.2.2. Solving $SD_2\{S_1(\hat{\Phi}, \delta_g)\} = \mathbf{0}$ for $\hat{\Phi}$ leads to a total of $p - 1$ systems of linear equations. One of these systems is 1×1 and the solution is $\hat{\varphi}_{21} = -\kappa^{21}/\kappa^{11}$. Solving the remaining systems gives the updates

$$
\begin{pmatrix} \hat{\varphi}_{r,r-2} \\ \hat{\varphi}_{r,r-1} \end{pmatrix} = - \begin{pmatrix} \kappa^{r-2,r-2} & \kappa^{r-1,r-2} \\ \kappa^{r-2,r-1} & \kappa^{r-1,r-1} \end{pmatrix}^{-1} \begin{pmatrix} \kappa^{r,r-2} \\ \kappa^{r,r-1} \end{pmatrix},
$$

for $r = 3, \ldots, p$. These estimates can be extended to the E_dVA case as follows:

$$
\begin{pmatrix} \hat{\varphi}^{(g)}_{r,r-d} \\ \hat{\varphi}^{(g)}_{r,r-(d-1)} \\ \vdots \\ \hat{\varphi}^{(g)}_{r,r-1} \end{pmatrix} =
$$
$$
- \begin{pmatrix} \kappa^{r-d,r-d} & \kappa^{r-(d-1),r-d} & \cdots & \kappa^{r-1,r-d} \\ \kappa^{r-d,r-(d-1)} & \kappa^{r-(d-1),r-(d-1)} & \cdots & \kappa^{r-1,r-(d-1)} \\ \vdots & \vdots & \ddots & \vdots \\ \kappa^{r-d,r-1} & \kappa^{r-(d-1),r-1} & \cdots & \kappa^{r-1,r-1} \end{pmatrix}^{-1} \begin{pmatrix} \kappa^{r,r-d} \\ \kappa^{r,r-(d-1)} \\ \vdots \\ \kappa^{r,r-1} \end{pmatrix},
$$

for $r = 2, \ldots, p$. Now, solving $S_2(\hat{\mathbf{T}}, \hat{\delta}_g) = 0$ gives the result in (8.15).

8.2.4 Modelling the Component Means

McNicholas and Subedi (2012) consider a linear model for the component means, while also using the covariance structures in Table 8.1. To facilitate modelling $\boldsymbol{\mu}_g$ using a line with intercept a_g and slope b_g, they define

$$
\mathbf{Q} = \begin{pmatrix} 1 & 1 & \cdots & 1 \\ t_1 & t_2 & \cdots & t_p \end{pmatrix}' \quad \text{and} \quad \boldsymbol{\beta}_g = \begin{pmatrix} a_g \\ b_g \end{pmatrix}.
$$

The likelihood for the VEA model in this case can be written

$$
\mathcal{L}(\boldsymbol{\vartheta}) = \prod_{i=1}^{n} \sum_{g=1}^{G} \pi_g \phi(\mathbf{x}_i \mid \mathbf{Q}\boldsymbol{\beta}_g, (\mathbf{T}'_g \mathbf{D}^{-1} \mathbf{T}_g)^{-1}). \tag{8.16}
$$

Once again, the EM algorithm can be used for parameter estimation. Another objective of the work of McNicholas and Subedi (2012) was to move from Gaussian to t-mixtures and this will be explored in Section 8.3.

In terms of modelling the mean, the approach of McNicholas and Subedi (2012) can easily be extended to other situations if desired. For example, in some situations it might be desirable to model the means using quadratic curves, in which case one may define

$$\mathbf{Q} = \begin{pmatrix} 1 & 1 & \cdots & 1 \\ t_1 & t_2 & \cdots & t_p \\ t_1^2 & t_2^2 & \cdots & t_p^2 \end{pmatrix}' \qquad \text{and} \qquad \boldsymbol{\beta}_g = \begin{pmatrix} a_g \\ b_g \\ c_g \end{pmatrix}.$$

A similar approach could be taken to a cubic or other polynomial model for the mean.

8.3   Using t-Mixtures

McNicholas and Subedi (2012) consider a t-analogue of the family of mixture models discussed in Section 8.2. They develop mixtures of multivariate t-distributions with component scale matrices decomposed as in Table 8.1, the option of a linear model for the mean (Section 8.2.4), and the option to constrain degrees of freedom to be equal across groups. Consider the VEA model with a linear mean and no constraints on the degrees of freedom. The likelihood for p-dimensional $\mathbf{x}_1, \ldots, \mathbf{x}_n$ from this model is given by

$$\mathcal{L}(\boldsymbol{\vartheta}) = \prod_{i=1}^{n} \sum_{g=1}^{G} \pi_g f(\mathbf{x}_i \mid \mathbf{Q}\boldsymbol{\beta}_g, (\mathbf{T}_g' \mathbf{D}^{-1} \mathbf{T}_g)^{-1}, \nu_g), \qquad (8.17)$$

where $f(\mathbf{x}_i \mid \mathbf{Q}\boldsymbol{\beta}_g, (\mathbf{T}_g' \mathbf{D}_g^{-1} \mathbf{T}_g)^{-1}, \nu_g)$ is the density of a p-dimensional t-distributed random vector with scale matrix $\boldsymbol{\Sigma}_g = (\mathbf{T}_g' \mathbf{D}^{-1} \mathbf{T}_g)^{-1}$, ν_g degrees of freedom, and mean

$$\boldsymbol{\mu}_g = \mathbf{Q}\boldsymbol{\beta}_g = \begin{pmatrix} 1 & 1 & \cdots & 1 \\ t_1 & t_2 & \cdots & t_p \end{pmatrix}' \begin{pmatrix} a_g \\ b_g \end{pmatrix},$$

for $a_g, b_g \in \mathbb{R}$. As in the Gaussian case, an EM algorithm can be used for parameter estimation. Proceeding as outlined in Section 5.1, a latent variable W_{ig} is introduced so that $\mathbf{X}_i \mid w_{ig}, z_{ig} = 1 \sim \mathrm{N}(\mathbf{Q}\boldsymbol{\beta}_g, (\mathbf{T}_g' \mathbf{D}^{-1} \mathbf{T}_g)^{-1}/w_{ig})$, independently, for $i = 1, \ldots, n$, and $W_{ig} \mid z_{ig} = 1$ follows a gamma distribution with parameters $(\nu_g/2, \nu_g/2)$, independently (cf. Section 5.1). It follows that the complete-data consist of the observed $\mathbf{x}_1, \ldots, \mathbf{x}_n$ together with the unobserved labels $\mathbf{z}_1, \ldots, \mathbf{z}_n$ and the latent w_{ig}, for $i = 1, \ldots, n$ and $g = 1, \ldots, G$.

Following McNicholas and Subedi (2012), the complete-data log-likelihood can be written

$$l_c(\boldsymbol{\vartheta}) = l_{1c}(\boldsymbol{\pi}) + l_{2c}(\boldsymbol{\nu}) + l_{3c}(\boldsymbol{\varphi}), \qquad (8.18)$$

where $\boldsymbol{\varphi} = (\boldsymbol{\beta}_1, \ldots, \boldsymbol{\beta}_G, \mathbf{T}_1, \ldots, \mathbf{T}_G, \mathbf{D}_1, \ldots, \mathbf{D}_G)$, $\boldsymbol{\pi} = (\pi_1, \ldots, \pi_G)$, and $\boldsymbol{\nu} =$

$(\nu_1, \ldots, \nu_G)$. The constituent elements of (8.18) are given by

$$l_{1c}(\boldsymbol{\pi}) = \sum_{g=1}^{G} n_g \log \pi_g,$$

$$l_{2c}(\boldsymbol{\nu}) = \sum_{g=1}^{G} \sum_{i=1}^{n} z_{ig} \left[-\log \Gamma \left(\frac{\nu_g}{2} \right) + \frac{\nu_g}{2} \log \left(\frac{\nu_g}{2} \right) \right.$$
$$\left. + \frac{\nu_g}{2} (\log w_{ig} - w_{ig}) - \log w_{ig} \right],$$

$$l_{3c}(\boldsymbol{\varphi}) = \frac{np}{2} \log(2\pi) - \sum_{g=1}^{G} \frac{n_g}{2} \log |\mathbf{D}| - \sum_{g=1}^{G} \frac{n_g}{2} \operatorname{tr}\{\mathbf{T}_g \mathbf{S}_g \mathbf{T}_g' \mathbf{D}^{-1}\},$$

where $n_g = \sum_{i=1}^{n} z_{ig}$ and

$$\mathbf{S}_g = \frac{1}{n_g} \sum_{i=1}^{n} z_{ig} w_{ig} (\mathbf{x}_i - \mathbf{Q}\boldsymbol{\beta}_g)(\mathbf{x}_i - \mathbf{Q}\boldsymbol{\beta}_g)'. \tag{8.19}$$

At each E-step, the missing data, z_{ig} and w_{ig}, are replaced by their conditional expected values,

$$\hat{z}_{ig} := \mathbb{E}[Z_{ig} \mid \mathbf{x}_i] = \frac{\hat{\pi}_g f(\mathbf{x}_i \mid \mathbf{Q}\hat{\boldsymbol{\beta}}_g, (\hat{\mathbf{T}}_g' \hat{\mathbf{D}}^{-1} \hat{\mathbf{T}}_g)^{-1}, \hat{\nu}_g)}{\sum_{h=1}^{G} \hat{\pi}_h f(\mathbf{x}_i \mid \mathbf{Q}\hat{\boldsymbol{\beta}}_g, (\hat{\mathbf{T}}_g' \hat{\mathbf{D}}^{-1} \hat{\mathbf{T}}_g)^{-1}, \hat{\nu}_g)} \tag{8.20}$$

and

$$\hat{w}_{ig} := \mathbb{E}[W_{ig} \mid \mathbf{x}_i, z_{ig} = 1] = \frac{\hat{\nu}_g + p}{\hat{\nu}_g + \delta(\mathbf{x}_i, \mathbf{Q}\hat{\boldsymbol{\beta}}_g \mid (\hat{\mathbf{T}}_g' \hat{\mathbf{D}}^{-1} \hat{\mathbf{T}}_g)^{-1})}, \tag{8.21}$$

respectively. As usual, expectations are conditional on current parameter estimates. In the M-step, and from $l_{1c}(\boldsymbol{\pi})$, the mixing proportions are updated by $\hat{\pi}_g = n_g/n$ and, from $l_{2c}(\boldsymbol{\nu})$, the estimates for the ν_g can be found by solving the equation

$$1 - \varphi \left(\frac{\hat{\nu}_g^{\text{new}}}{2} \right) + \frac{1}{n_g} \sum_{i=1}^{n} \hat{z}_{ig} (\log \hat{w}_{ig} - \hat{w}_{ig}) + \log \left(\frac{\hat{\nu}_g^{\text{new}}}{2} \right)$$
$$+ \varrho \left(\frac{\hat{\nu}_g + p}{2} \right) - \log \left(\frac{\hat{\nu}_g + p}{2} \right) = 0, \tag{8.22}$$

for $\hat{\nu}_g^{\text{new}}$, where $\hat{\nu}_g$ is the estimate of ν_g from the previous iteration and $\varrho(\cdot)$ is the digamma function. From $l_{3c}(\boldsymbol{\varphi})$, the update for $\boldsymbol{\beta}_g$ is given by

$$\hat{\boldsymbol{\beta}}_g = (\mathbf{Q}'\mathbf{T}_g' \mathbf{D}^{-1} \mathbf{T}_g \mathbf{Q})^{-1} \mathbf{Q}'\mathbf{T}_g' \mathbf{D}^{-1} \mathbf{T}_g \frac{\sum_{i=1}^{n} \hat{z}_{ig} \hat{w}_{ig} \mathbf{x}_i}{\sum_{j=1}^{n} \hat{z}_{jg} \hat{w}_{jg}}.$$

As one would expect, if the linear constraint is not imposed on the mean, then the update is just

$$\hat{\boldsymbol{\mu}}_g = \frac{\sum_{i=1}^{n} \hat{z}_{ig} \hat{w}_{ig} \mathbf{x}_i}{\sum_{j=1}^{n} \hat{z}_{jg} \hat{w}_{jg}}.$$

The estimates for $\mathbf{T}_g$ and $\mathbf{D}$ are as given in Section 8.2.2.1 but with $\mathbf{S}_g$ defined as in (8.19). Details are given by McNicholas and Subedi (2012).

8.4   Illustrations

8.4.1   Clustering

Simulated data are used to illustrate the model-based clustering of longitudinal data. A simple approach is taken to data generation, where the first step is to generate one representative time course for each component (Figure 8.1). Then other time courses arise around these representative time courses by adding random quantities from a uniform$(-u, u)$ distribution. For smaller values of u, the result is a very easy clustering problem, e.g., $u = 0.5$, with 100 expressions per component, as shown in Figure 8.2.

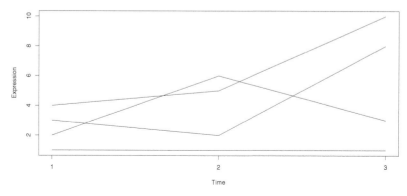

Figure 8.1
Four representative time courses.

Repeating the data generation process with $u = 1.5$ leads to a somewhat more difficult clustering problem (Figure 8.2), which will serve as a good first clustering example. Using the `longclust` package, each member of the CDGMM family is fitted to these data for $G = 2, \ldots, 6$ using one random start. The BIC selects a $G = 4$ component EEI model, and the associated MAP classifications correspond to perfect class agreement (Figure 8.3). Repeating the process for the CDtMM family (i.e., the t-analogue of the CDGMM family) leads to a $G = 4$ component EEA model with perfect classifications. Note that

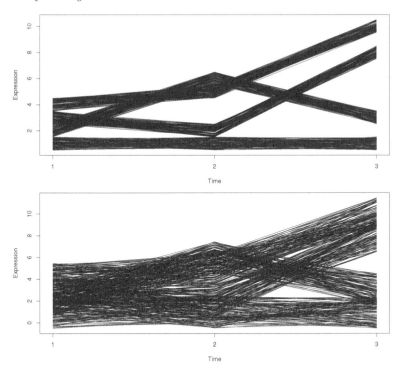

Figure 8.2
Two simulated longitudinal clustering problems with $n = 400$, three time points, and four clusters: one is very easy ($u = 0.5$; top) and the other is somewhat more difficult ($u = 1.5$; bottom).

the degrees of freedom for each component are high, reflecting a fit that is very close to a Gaussian mixture. Note also that some of the CDtMM models cannot be fitted from the (one) random start used by `longclust`.

A similar data generation process is used to generate a dataset with four clusters and five time points. This is a more challenging clustering problem (Figure 8.4). Again, each member of the CDGMM family is fitted to these data for $G = 2, \ldots, 6$ using one random start. The BIC selects a $G = 4$ component EEI model, and the associated MAP classifications correspond to almost perfect class agreement (ARI = 0.99; Figure 8.4). Repeating the process for the CDtMM family leads to a $G = 4$ component EEA model with the same classifications. The degrees of freedom for each component are high, reflecting a fit that is very close to a Gaussian mixture, i.e., $\hat{\nu} = (219.84, 215.84, 218.81, 216.21)$. However, the output from `longclust` suggests that most of the CDtMM models cannot be fitted from the (one) random start used. Repeating the CDtMM analysis from ten random starts results in a wider variety of models being fitted, but the best model is still a $G = 4$ component EEA model with ARI = 0.99.

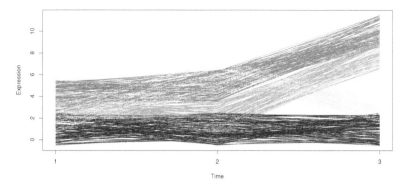

Figure 8.3
Time courses for the simulated longitudinal clustering problem with $u = 1.5$, $n = 400$, three time points, and four clusters, where shading reflects predicted classifications.

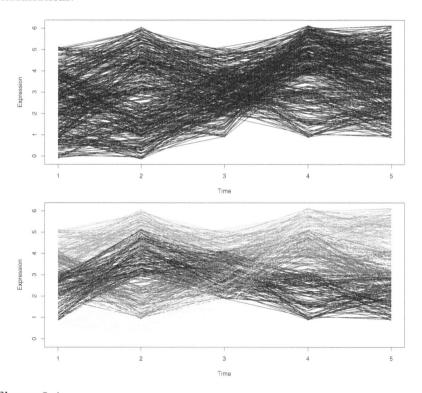

Figure 8.4
Time courses for simulated longitudinal data with five time points (top) and the same time courses shaded by predicted classifications (bottom).

Finally, it is useful to consider a difficult longitudinal clustering example. Following an analogous data generation process as before, a dataset with three clusters and eight time points is generated (Figure 8.5).

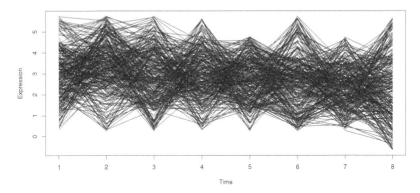

Figure 8.5
Time courses for a difficult clustering problem, i.e., simulated longitudinal data with eight time points and $u = 1.75$.

Again, each member of the CDGMM family is fitted to these data for $G = 2, \ldots, 6$ using ten random starts. The BIC selects a $G = 3$ component EEI model, and the associated MAP classifications correspond to good class agreement (ARI $= 0.83$; Table 8.2). Repeating the process for the CDtMM family leads to a $G = 3$ component EVI model with similar classifications (ARI $= 0.83$; Table 8.3). The degrees of freedom for each component are again high, reflecting a fit that is very close to a Gaussian mixture. The CDtMM solution can be visualized as in Figure 8.6.

Table 8.2
Cross-tabulation of the MAP classifications (A–C) associated with the selected CDGMM, using ten random starts, against true classes for the eight-time-point longitudinal data.

	A	**B**	**C**
1	98	1	1
2	2	88	10
3	0	4	96

8.4.2 Classification

Continuing in the same vein as Section 8.4.1, data can be generated that present a challenging classification problem and a practically impossible clus-

Table 8.3
Cross-tabulation of the MAP classifications (A–C) associated with the selected
CtGMM, using ten random starts, against true classes for the eight-time-point
longitudinal data.

	A	B	C
1	99	0	1
2	2	87	11
3	0	4	96

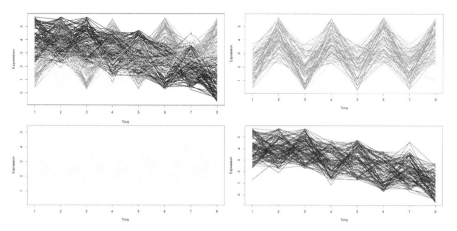

Figure 8.6
Time courses for a difficult clustering problem, with shading corresponding
to MAP classifications from the selected CDtMM model, shown on the same
plot (top left) and separately.

tering problem. Consider the time courses in Figure 8.7. Not surprisingly,
applying the CDGMM and CDtMM families to these data for clustering leads
to poor results. Specifically, using 20 random starts, the BIC selects a $G = 2$
component EEI model with ARI = 0.37. The selected CDtMM model is a
little better: a $G = 3$ component EEI model with ARI = 0.48.

While too difficult to be an effective illustration of clustering, these data
can be used to demonstrate model-based classification of longitudinal data.
A classification scenario is simulated by taking every fourth time course to
be unlabelled, with the effect that precisely 75 of the 300 time courses are
unlabelled. All CDGMM models are run for $G = 3$ and the BIC selects an
EEI model that gives reasonable classification performance (ARI = 0.52, Table 8.4). Running all CDtMM models for $G = 3$, the BIC selects an EEI model
that gives slightly better classification performance (ARI = 0.54, Table 8.5).

Based on the same dataset, another classification scenario is simulated

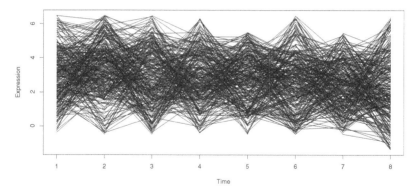

Figure 8.7
Time courses for a difficult classification problem, i.e., simulated longitudinal data with eight time points and $u = 2.5$.

Table 8.4
Cross-tabulation of the MAP classifications (A–C) associated with the selected CDGMM model against true classes for simulated time course data, where every fourth time course is treated as unlabelled.

	A	B	C
1	24	0	1
2	5	17	3
3	0	6	19

Table 8.5
Cross-tabulation of the MAP classifications (A–C) associated with the selected CDtMM model against true classes for simulated time course data, where every fourth time course is treated as unlabelled.

	A	B	C
1	24	0	1
2	5	17	3
3	0	5	20

by taking every fourth time course to be unlabelled, starting with the third time course (rather than the fourth). Again, the effect is that precisely 75 of the 300 time courses are unlabelled. All CDGMM models are run for $G = 3$ and the BIC selects an EEI model that gives good classification performance

(ARI = 0.67, Table 8.6). Running all CD*t*MM models for $G = 3$, the BIC selects an EEI model that gives identical classification performance to the selected CDGMM; the selected CD*t*MM is effectively a Gaussian mixture model because all elements of $\hat{\boldsymbol{\nu}}$ are very large.

Table 8.6
Cross-tabulation of the MAP classifications (A–C) associated with the selected CDGMM model against true classes for simulated time course data (alternative partition), where one-quarter of the time courses are treated as unlabelled.

	A	B	C
1	23	0	2
2	1	21	3
3	0	3	22

8.5   Comments

Although illustrated on simulated data, the approaches discussed in this chapter have been used effectively in real data analyses (e.g., Humbert et al., 2013). Anderlucci and Viroli (2015) have extended the methodology of McNicholas and Murphy (2010a) to the situation where there are multiple responses for each individual at each time point. Their approach is nicely illustrated with data from a health and retirement study. The notion of constraining subdiagonals of $\mathbf{T}_g$ (cf. Section 8.2.3) deserves some further attention, within both the single- and multiple-response paradigms. Finally, it will be interesting to explore the use of mixtures of MPE distributions as an alternative to *t*-mixtures; whereas *t*-mixtures essentially allow more dispersion about the mean when compared with Gaussian mixtures, mixtures of MPE distributions would allow both more and less dispersion (cf. Chapter 5).

9

Miscellania

9.1   On the Definition of a Cluster

Intuition suggests that a cluster should comprise points that diffuse from a mode, and experience gained working on real problems reinforces this intuition. The author of the present monograph ("the author" hereafter) takes the view that a cluster is only believable if it is a component within a finite mixture model. Furthermore, the component should be unimodal and the mixture model should be appropriate in light of the data. Why unimodal? If the component is not unimodal, then one of two things is almost certainly happening: the wrong mixture distribution is being fitted (e.g., multiple Gaussian components being used to model one skewed cluster) or not enough components are being used. Of course, neither case is satisfactory. What does it mean for a mixture model to be "appropriate" in light of the data? It means that the model has the necessary flexibility, or parameterization, to fit the data; e.g., if the data contain skewed clusters, then the mixture model should be able to accommodate skewed components. In many cases, being appropriate in light of the data will also mean that each component has convex contours so that each cluster is convex (cf. Section 7.6).

The position taken here ties in with two of the definitions given when the historical development of the field was discussed in Section 2.1. Certainly, this position is compatible with the definition Tiedeman (1955) and Wolfe (1963) use for a cluster, i.e., that a cluster is one of the components of a mixture model, albeit it also contains elements of mode-related definitions such as that used by Cattell (1957), i.e., that a cluster is a mode within a distribution. The definition proffered herein can be stated succinctly as follows:

> A cluster is a unimodal component within an appropriate finite mixture model.

Again, an "appropriate" mixture model here is one that is appropriate in light of the data under consideration. That this definition ties the notion of a cluster to the data under consideration is essential because a cluster really only has meaning in the context of data. While this definition insists that clusters are unimodal, it is not at all the same as asserting that a cluster is a mode. The principal problem with defining a cluster in terms of a mode can be seen by generating two overlapping Gaussian components such that there are clearly three modes, e.g., Figure 9.1. This point is reinforced in Figure 9.2, which, similar to Figure 9.1, is produced using the `ggplot2` package (Wickham, 2009). Of course, applying the GPCM family to these data — or the *t*EIGEN family, the PGMM family, etc. — results in a two-component model with sensible associated classifications.

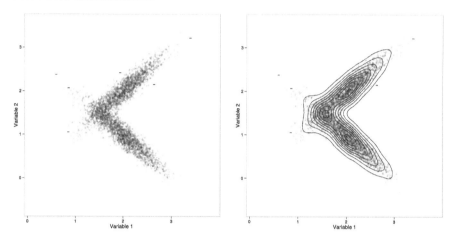

Figure 9.1
Scatter plots, one with contours, with semi-transparent points for data simulated from two overlapping Gaussian components.

The views thus far expounded in this section tie in with some of the arguments that have been proffered in favour of model-based clustering as a method, and these will be discussed in Section 9.2. Gordon (1981, Section 1.1) reports two desiderata, or desired characteristics, of a cluster that are stated as "basic ideas" by Cormack (1971):

> Two possible desiderata for a cluster can thus be stated as internal cohesion and external isolation.

Of course, complete external isolation will not be possible in many real analyses; however, the idea of internal cohesion seems quite compatible with the idea of a cluster corresponding to a unimodal component in an appropriate finite mixture. Interestingly, when referring to a situation where external isolation may not be possible, Gordon (1981, Sec. 1.1) highlights the fact that

> . . . the conclusion reached will in general depend on the nature of the data.

This vital link with the data under consideration is along similar lines to the requirement of an "appropriate" finite mixture model in the definition proffered herein.

Everitt et al. (2011, Section 1.4) point out that dissection, as opposed

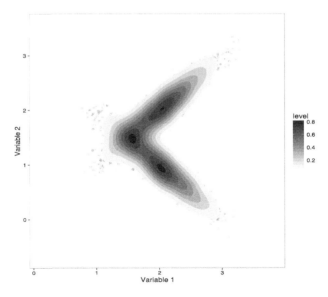

Figure 9.2
Visualization of the density of the scatter plot in Figure 9.1.

to clustering, might be necessary in some circumstances, and Gordon (1981, Section 1.1) makes a similar point. Everitt et al. (2011, Section 1.4) define dissection as

> ... the process of dividing a homogenous data set into different parts.

Of course, it is true that there are situations where one might wish to carry out dissection rather than clustering. In fact, there may even be cases where a departure from the definition of a cluster offered herein is desirable in light of the data under consideration. In general, however, the author does not feel comfortable reporting clustering results to scientists, or other collaborators, unless the clusters can be framed in terms of the (unimodal) components of an appropriate mixture model.

9.2 What Is the Best Way to Perform Clustering, Classification, and Discriminant Analysis?

At the end of his brilliant book, Synge (1951) writes:

> Time and again science has progressed by recognizing that what seems hard is actually easy and that what seems easy is actually hard.

Whether clustering, or statistics in general, is science may be debatable but those waters will not be navigated herein. The spirit of this quote from Synge (1951) is, however, highly applicable to clustering. Although sometimes initially dismissed as easy, the process of carrying out a cluster analysis on real data is generally quite hard. Depending on certain factors, e.g., the proportion of labelled observations, the same is true to a greater or lesser extent for semi-supervised classification. Recognizing this inherent difficulty is crucial in the context of the discussion in this section.

The answer to the question posed in the title of this section, from the point of view of the author, is that there is no single best method for clustering, classification, or discriminant analysis. Because it is a supervised approach, determining the best discriminant analysis approach for a given dataset is often relatively straightforward, especially when the proportion of labelled observations is high. However, as the model-based discriminant analysis of the olive oil data in Section 2.6.4 highlighted, one can observe very good classification results for the training set and yet obtain disappointing classification results on the test set. Notably, more than one component per class was used in this analysis, and discriminant analyses of the same data, where one (more flexible/appropriate) component per class was used in each case, yielded excellent classification performance (Section 7.7.3). In general, while mixture model-based approaches can be effective for discriminant analysis, great care should be taken if more than one component per class is used; rather, the author finds it preferable to use one appropriate density per class, in which case the procedure is essentially straightforward discriminant analysis. These comments conclude the discussion on discriminant analysis, and the discussion now focuses on clustering and (semi-supervised) classification.

For quite some time, there has been support for the model-based approach to clustering. McLachlan and Peel (2000a, Section 1.15.1) cite some of the better-known proponents of model-based clustering. This includes Aitkin et al. (1981), who write that

> ...when clustering samples from a population, no cluster method is *a priori* believable without a statistical model.

McLachlan and Peel (2000a, Section 1.15.1) also cite Marriott (1974), who writes that the model-based approach

> ...is about the only clustering technique that is entirely satisfactory from a mathematical point of view. It assumes a well-defined mathematical model, investigates it by well-established statistical techniques, and provides a test of significance for the results.

If one accepts the definition of a cluster given herein (cf. Section 9.1), then it makes sense to use mixture model-based approaches for clustering and, by extension, for (semi-supervised) classification. However, it is not necessarily true that other approaches will not perform as well, or even better, on a given dataset. For example, Steinley and Brusco (2011a) discuss the performance of

k-means clustering versus Gaussian model-based clustering and argue that the former approach might be preferable in some circumstances. In a commentary on this work, Vermunt (2011) illustrates that k-means clustering is equivalent to the EII GPCM model with a classification EM algorithm used for parameter estimation (cf. Celeux and Govaert, 1992). As illustrated herein, k-means clustering can be very useful for, *inter alia*, initializing model-based clustering approaches.

In a real cluster analysis, it is recommended that several approaches be used and the results compared. Comparing results from different model-based clustering approaches is greatly aided by the fact that model-based clustering methods return soft classifications, i.e., the *a posteriori* $\hat{z}_{ig}$. This allows a detailed study of the observations on which different model-based methods disagree. Making such a comparison between approaches that cannot return soft classifications is more problematic. Accordingly, while comparing results from model-based approaches to one another for a given dataset is quite straightforward, comparing results from multiple non-model-based approaches or comparing model-based and non-model-based approaches can be difficult (unless the non-model-based approaches also return soft classifications). When using mixture model-based approaches, one might suggest that only the most general models are needed. For instance, it might be posited that Gaussian mixtures have been superseded by t-mixtures or that the very existence of the MGHD renders useless its special and limiting cases. From various illustrations herein, it is clear that the more flexible model is not necessarily better for a given dataset. In fact, whether a single best model should be chosen at all is debatable and this will be discussed further, in the mixture model-based context, in Section 9.3.

Because this section concerns discussion of the best clustering or classification method, it makes sense to consider how a newly introduced clustering method might be evaluated in the context of existing approaches. When trying to convince a reader that a new clustering method is useful over and above existing methods, it is almost unavoidable to use, for illustration, examples where the underlying classes are known. Certain datasets have become popular within the model-based clustering literature. The more a dataset is used, the more attractive it is for comparisons because clustering results for many methods are known and so comparison with several approaches is readily available — this is essential in a field where effective software is often not freely available. One such dataset is the Italian wine dataset from the `gclus` package for R; it has been used in several papers (e.g., McNicholas and Murphy, 2008; Scrucca, 2010) and its popularity is understandable because it presents the opportunity to show that a new method beats results reported for several other approaches. Would a superior result mean that the new method is better than all of those other approaches? No, of course not. Provided that the analyses are really comparable, it means that the new method works better for at least one dataset, with particular starting values, etc. This is, of course, useful information but there is some inherent risk in putting too much stock in whether a new method beats other methods for particular datasets.

In general, if a clustering method is shown to perform very well on a given dataset such that it outperforms other methods that also perform very well on that dataset, then it is a useful addition to the suite, or the toolbox, of clustering methods. However, this is not the only way that a clustering method can be shown to be useful; e.g., it might be effective at narrowing down solutions or for helping to verify results from other methods. Also, a method may perform well on several datasets and yet not be the best method on any one; such methods can also be useful, *inter alia*, for finding starting values for other methods. The notion that there are certain datasets against which clustering methods should be benchmarked perhaps misses the point. The most obvious danger is that such benchmarking may lead to the dismissal of new approaches that would perform very well on datasets other than the benchmark datasets. Another danger is that, if there is a set of accepted benchmark datasets, then methodological development may become tailored to these datasets. There is also the aforementioned fact that a clustering method that is never the best one when applied to these datasets may still be useful.

9.3   Mixture Model Averaging

When families of mixture models are used, the typical procedure is to fit many models and select the "best" one based on some criteria, often the BIC. Then one proceeds as if the selected model has generated the data, or at least provides a good representation thereof, without considering the other models that were fitted. From a statistical inference viewpoint, one can argue that ignoring the other models is problematic. In practice, it is unclear what should be done when the difference between the values of the criteria for two different models is "small". Model averaging takes model uncertainty into consideration by combining parameter estimates across different models. Bayesian model averaging (BMA; Hoeting et al., 1999) is a popular approach and is used in model-based clustering applications of the GPCM family by Wei and McNicholas (2015). Given the popularity of the BIC in model-based applications, BMA is a natural choice.

Using the notation of Hoeting et al. (1999), let Δ be the quantity of interest and consider models $\mathcal{M}_1, \mathcal{M}_2, \ldots, \mathcal{M}_K$. The posterior distribution of Δ given data $\mathbf{x}$ is

$$\mathrm{pr}(\Delta \mid \mathbf{x}) = \sum_{i=1}^{K} \mathrm{pr}(\Delta \mid \mathcal{M}_i, \mathbf{x}) \mathrm{pr}(\mathcal{M}_i \mid \mathbf{x}), \qquad (9.1)$$

where $\mathrm{pr}(\Delta \mid \mathcal{M}_i, \mathbf{x})$ is the posterior distribution for Δ under model $\mathcal{M}_i$,

$\mathrm{pr}(\mathcal{M}_i \mid \mathbf{x})$ is the posterior probability for model $\mathcal{M}_i$, i.e.,

$$\mathrm{pr}(\mathcal{M}_i \mid \mathbf{x}) = \frac{\mathrm{pr}(\mathbf{x} \mid \mathcal{M}_i)\mathrm{pr}(\mathcal{M}_i)}{\sum_{k=1}^{K} \mathrm{pr}(\mathbf{x} \mid \mathcal{M}_k)\mathrm{pr}(\mathcal{M}_k)}, \tag{9.2}$$

and

$$\mathrm{pr}(\mathbf{x} \mid \mathcal{M}_i) = \int \mathrm{pr}(\mathbf{x} \mid \boldsymbol{\theta}_i, \mathcal{M}_i)\mathrm{pr}(\boldsymbol{\theta}_i \mid \mathcal{M}_i)d\boldsymbol{\theta}_i, \tag{9.3}$$

where $\boldsymbol{\theta}_i$ is the vector of parameters for model $\mathcal{M}_i$, $\mathrm{pr}(\boldsymbol{\theta}_i \mid \mathcal{M}_i)$ is the prior distribution for $\boldsymbol{\theta}_i$ under model $\mathcal{M}_i$, and $\mathrm{pr}(\mathcal{M}_i)$ is the prior probability of model $\mathcal{M}_i$. There are some practical difficulties with the computations involved in BMA, e.g., the number of models in the summation in (9.1) can be prohibitively large, and so Madigan and Raftery (1994) propose using Occam's window to choose a set of models. That is, models not in Occam's window

$$\left\{ \mathcal{M}_i : \frac{\max_l \{\mathrm{pr}(\mathcal{M}_l \mid \mathbf{x})\}}{\mathrm{pr}(\mathcal{M}_i \mid \mathbf{x})} \leq c \right\} \tag{9.4}$$

are discarded, where c is some positive number. Madigan and Raftery (1994) use $c = 20$ in their analyses, based on an analogy with a p-value of 0.05.

In mixture model applications, the BIC can be used to approximate the integral in (9.3) (Dasgupta and Raftery, 1998). Specifically,

$$\mathrm{pr}(\mathbf{x} \mid \mathcal{M}_i) \approx \exp\left\{\frac{1}{2}\mathrm{BIC}_i\right\},$$

where BIC_i is the value of the BIC for model $\mathcal{M}_i$, cf. (2.17). Therefore, assuming equal prior probabilities, (9.2) can be written

$$\mathrm{pr}(\mathcal{M}_i \mid \mathbf{x}) \approx \frac{\exp\left\{\frac{1}{2}\mathrm{BIC}_i\right\}}{\sum_{k=1}^{K} \exp\left\{\frac{1}{2}\mathrm{BIC}_k\right\}}, \tag{9.5}$$

and Occam's window (9.4) is approximately equivalent to

$$\left\{ \mathcal{M}_i : \max_l\{\mathrm{BIC}_l\} - \mathrm{BIC}_i \leq 2\log c \right\}. \tag{9.6}$$

Wei and McNicholas (2015) use (9.5) to compute weights for model-based clustering averaging approaches, and Occam's window (9.6) to decide which models to average. Before describing the averaging approaches used by Wei and McNicholas (2015), it is necessary to review their approach for merging mixture components — which is necessary when models in Occam's window have different numbers of components and we want to use all of the models. Consider a G-component mixture model and suppose the objective is to merge components to produce an H-component mixture, where $H < G$. The density of the resulting mixture is just an alternative representation of the original model, i.e.,

$$f(\mathbf{x}) = \sum_{j=1}^{H} \pi_j^* f_j^*(\mathbf{x}) = \sum_{g=1}^{G} \pi_g \phi(\mathbf{x} \mid \boldsymbol{\mu}_g, \boldsymbol{\Sigma}_g), \tag{9.7}$$

where each π_j^* is equal to one of or the sum of some of the mixing proportions $\pi_1, \ldots, \pi_G$, and each $f_j^*(\mathbf{x})$ is one of or a mixture of some of the component densities $\phi(\mathbf{x} \mid \boldsymbol{\mu}_1, \boldsymbol{\Sigma}_1), \ldots, \phi(\mathbf{x} \mid \boldsymbol{\mu}_G, \boldsymbol{\Sigma}_G)$. Wei and McNicholas (2015) develop a mixture model component merging criterion based on the ARI with respect to a "reference model". The purpose of merging components is that models in Occam's window have the same number of components as the reference model so that all can be used if desired. Two cases are considered.

- In Case I, the reference model is the model with the largest BIC and merging is carried out for models in Occam's window with more components than the reference model. Models in Occam's window with fewer components than the reference model are discarded.

- In Case II, the reference model is the model with the fewest components and merging is performed for models in Occam's window with more components.

Of course, Cases I and II are equivalent when the model with the largest BIC has the fewest components. In Case I, there is the implicit assumption that the BIC will not underestimate the number of components; however, this assumption is not made in Case II.

Wei and McNicholas (2015) present two approaches for averaging the models in Occam's window. One approach is to compute a weighted average of the *a posteriori* probabilities $\hat{z}_{ig}$ for the models in Occam's window, where the weights are computed using (9.5). Another approach is to average the models in Occam's window by taking a weighted average of the parameter estimates for each parameter; again, the weights are computed using (9.5). Extensive details, along with examples, are given by Wei and McNicholas (2015).

9.4   Robust Clustering

In real applications, one may encounter data that are contaminated by outliers, noise, or generally spurious points. Borrowing the terminology used by Aitkin and Wilson (1980), these types of observations can be collectively referred to as "bad" while all others can be called "good". When bad points are present, they can have a deleterious effect on mixture model parameter estimation. Accordingly, it is generally desirable to account for bad points when present. One way to do this is to use a mixture of distributions with component concentration parameters. Many such mixtures have already been considered herein and include t-mixtures and power exponential mixtures; however, Hennig (2004) points out that t-mixtures are vulnerable to "very extreme outliers" and the same is probably true for robustness-via-component concentration parameter approaches in general.

Within the Gaussian mixture paradigm, Campbell (1984), McLachlan and Basford (1988), Kharin (1996), and De Veaux and Krieger (1990) use M-estimators (Huber, 1964, 1981) of the component means and covariance matrices. In a similar vein, Markatou (2000) utilizes a weighted likelihood approach to obtain robust parameter estimates. Banfield and Raftery (1993) add a uniform component on the convex hull of the data to accommodate outliers in a Gaussian mixture model, and Fraley and Raftery (1998) and Schroeter et al. (1998) further consider approaches in this direction. Hennig (2004) suggests adding an improper uniform distribution as an additional mixture component. Browne et al. (2012) also make use of uniform distributions but they do so by making each component a mixture of a Gaussian and a uniform distribution. Rather than specifically accommodating bad points, this approach allows for what they call "bursts" of probability as well as locally heavier tails — of course, this might have the effect of dealing with bad points for some datasets.

García-Escudero et al. (2008) outline a trimmed clustering approach that gives robust parameter estimates by allowing for a pre-specified proportion of bad points. They achieve this by imposing "eigenvalues-ratio restrictions" (see García-Escudero et al., 2008, for details). These constraints can be viewed as a multivariate extension of the univariate work of Hathaway (1985). The trimmed clustering approach of García-Escudero et al. (2008) has been applied for Gaussian mixtures and is implemented as such in the R package `tclust` (Fritz et al., 2012). The approach can be very effective when the number of variables p is sufficiently small that the proportion of bad points can be accurately pre-specified. Although work to date has somewhat focused on Gaussian mixtures, an analogous approach could be taken to mixtures with non-Gaussian components.

Punzo and McNicholas (2016) use a mixture of contaminated Gaussian distributions, with density of the form

$$f\left(\mathbf{x} \mid \boldsymbol{\vartheta}\right) = \sum_{g=1}^{G} \pi_g \left[\alpha_g \phi\left(\mathbf{x} \mid \boldsymbol{\mu}_g, \boldsymbol{\Sigma}_g\right) + (1 - \alpha_g) \phi\left(\mathbf{x} \mid \boldsymbol{\mu}_g, \eta_g \boldsymbol{\Sigma}_g\right)\right], \quad (9.8)$$

where $\alpha_g \in (0.5, 1)$ is the proportion of good points in the gth component and $\eta_g > 1$ is the degree of contamination. This approach is available as the `ContaminatedMixt` package (Punzo et al., 2015) for R. Because $\eta_g > 1$ is an inflation parameter, it can be interpreted as the increase in variability due to the bad observations (e.g., Figure 9.3). This contaminated Gaussian mixture approach, i.e., (9.8), can be viewed as a special case of the multi-layer mixture of Gaussian distributions of Li (2005), where each of the G components at the top layer is itself a mixture of two components, with equal means and proportional covariance matrices at the secondary layer.

One major advantage of the mixture of contaminated Gaussian distributions approach is that the proportion of bad points does not need to be specified *a priori* (cf. Punzo and McNicholas, 2016). As a result, it is possible to apply this approach to higher-dimensional data and even to high-dimensional

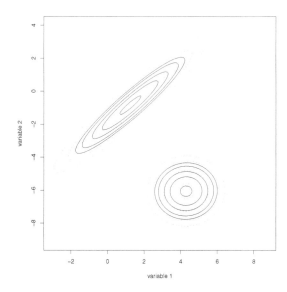

Figure 9.3
Contour plot depicting a two-component mixture of contaminated Gaussian distributions, where the contaminated parts are represented by broken lines.

data, e.g., via mixtures of contaminated Gaussian factor analyzers (Punzo and McNicholas, 2014b). Furthermore, this approach is easily extended to other cases, including non-elliptical components (Morris et al., 2014) and cluster-weighted models (Punzo and McNicholas, 2014a; cf. Section 9.6).

Ritter (2015) considers that bad points can be categorized as either mildly atypical or grossly atypical. It is worth noting that, while the contaminated mixture approach can work very well for mildly atypical points, a different approach is needed for grossly atypical points. Grossly atypical points are essentially what Hennig (2004) refers to as very extreme outliers, and trimming is usually required. This opens up the possibility of using trimming to remove grossly atypical points, which are relatively easy to detect, followed by a contaminated mixture approach to account for mildly atypical points.

9.5 Clustering Categorical Data

Latent class analysis has been widely used for clustering of categorical data and data of mixed type (e.g., Goodman, 1974; Celeux and Govaert, 1991; Biernacki et al., 2010). Much work on refinement and extension has been carried out. For example, Vermunt (2003, 2007) develop multilevel latent class models to account for conditional dependency between the response variables, and Marbac et al. (2014) propose a conditional modes model that assigns response variables into conditionally independent blocks. Besides latent class analysis, mixture model-based approaches for categorical data have received relatively little attention within the literature.

Browne and McNicholas (2012) develop a mixture of latent variables model for clustering of data with mixed type (cf. Section 9.7), and a dataset comprising only categorical (including binary) variables fits within their modelling framework as a special case. Browne and McNicholas (2012) draw on the deterministic annealing approach of Zhou and Lange (2010) in their parameter estimation scheme. This approach can increase the chance of finding the global maximum; however, Gauss–Hermite quadrature is required to approximate the likelihood. Gollini and Murphy (2014) use a mixture of latent trait analyzers (MLTA) model to cluster categorical data. They also apply their approach to binary data, where a categorical latent variable identifies clusters of observations and a latent trait is used to accommodate within-cluster dependency. A lower bound approximation to the log-likelihood is used, which is straightforward to implement and converges relatively quickly compared with other numerical approximations to the likelihood. A mixture of item response models (Muthen and Asparouhov, 2006; Vermunt, 2007) has very similar structure to the MLTA model; however, it is highly parameterized, uses a probit structure, and numerical integration is required to compute the likelihood. A similar approach is discussed by Cagnone and Viroli (2012), who use Gauss–Hermite quadrature to approximate the likelihood; they also assume a semi-parametric distributional form for the latent variables by adding extra parameters to the model. Repeatedly sampled binary data can be clustered using multilevel mixture item response models (Vermunt, 2007).

Tang et al. (2015) propose two mixtures of latent traits models with common slope parameters for model-based clustering of binary data. One is a general model that supposes that the dependence among the response variables within each observation is wholly explained by a low-dimensional continuous latent variable in each component. The other is specifically designed for repeatedly sampled data and supposes that the response function in each cluster is made up of two continuous latent variables by adding a blocking latent variable. Their proposed mixture of latent trait models with common slope parameters (MCLT) model is a categorical analogue of the MCFA model (cf. Section 3.5). The MCLT model allows for significant reduction in the number of free parameters when estimating the slope. Moreover, it permits a low-dimensional visual representation of the clusters, where posterior means of the continuous latent variables correspond to the manifest data.

Ranalli and Rocci (2016) develop an approach for clustering ordinal data. They use an underlying response variable approach, which treats ordinal variables as categorical realizations of underlying continuous variables (cf. Jöreskog, 1990). A pairwise likelihood approach is taken (cf. de Leon, 2005), and a pairwise EM algorithm is used for parameter estimation.

9.6   Cluster-Weighted Models

Consider data of the form $(\mathbf{x}_1, y_1), \ldots, (\mathbf{x}_n, y_n)$ so that each observation is a realization of the pair $(\mathbf{X}, Y)$ defined on some space Ω, where $Y \in \mathbb{R}$ is a response variable and $\mathbf{X} \in \mathbb{R}^p$ is a vector of covariates. Suppose that Ω can be partitioned into G groups, say $\Omega_1, \ldots, \Omega_G$, and let $p(\mathbf{x}, y)$ be the joint density of $(\mathbf{X}, Y)$. In general, the density of a cluster-weighted model (CWM) can be written

$$p(\mathbf{x}, y \mid \boldsymbol{\vartheta}) = \sum_{g=1}^{G} \pi_g p(y \mid \mathbf{x}, \boldsymbol{\theta}_g) p(\mathbf{x} \mid \boldsymbol{\Phi}_g),$$

where $\boldsymbol{\vartheta} = (\boldsymbol{\pi}, \boldsymbol{\theta}_1, \ldots, \boldsymbol{\theta}_G, \boldsymbol{\Phi}_1, \ldots, \boldsymbol{\Phi}_G)$ denotes the model parameters. More specifically, the density of a linear Gaussian CWM is

$$p(\mathbf{x}, y \mid \boldsymbol{\vartheta}) = \sum_{g=1}^{G} \pi_g \phi_1 \left(y \mid \beta_{0g} + \boldsymbol{\beta}'_{1g}\mathbf{x}, \sigma_g^2 \right) \phi_p \left(\mathbf{x} \mid \boldsymbol{\mu}_g, \boldsymbol{\Sigma}_g \right), \tag{9.9}$$

where $\beta_{0g} \in \mathbb{R}$ and $\boldsymbol{\beta}_{1g} \in \mathbb{R}^p$. As usual, $\phi_j(\cdot)$ is the density of a j-dimensional random variable from a Gaussian distribution and $\boldsymbol{\vartheta}$ denotes all model parameters. The linear Gaussian CWM in (9.9) has been studied by Gershenfeld (1997) and Schöner (2000). CWMs are burgeoning into a vibrant subfield of model-based clustering and classification. For example, Ingrassia et al. (2012) consider an extension to t-distribution that leads to the linear t-CWM. Ingrassia et al. (2014) introduce a family of 12 parsimonious linear t-CWMs, Punzo (2014) introduces the polynomial Gaussian CWM, and Punzo and Ingrassia (2015) propose CWMs for bivariate data of mixed-type. Ingrassia et al. (2015) consider CWMs with categorical responses and also consider identifiability under the assumption of Gaussian covariates.

In the mosaic of work about the use of mixture models for clustering and classification, CWMs have their place in applications with random covariates. Indeed, as distinct from finite mixtures of regressions (e.g., Leisch, 2004; Frühwirth–Schnatter, 2006), which are examples of mixture models with fixed covariates, CWMs allow for assignment dependence, i.e., the distribution of the covariates in each component can also be distinct. From a clustering and classification perspective, this implies that the covariates $\mathbf{X}$ can directly affect the clustering results — for most applications, this represents an advantage over the fixed covariates approach (Hennig, 2000).

Applying model (9.9) in high dimensions is infeasible for the same reason that using the Gaussian mixture model with unconstrained $\boldsymbol{\Sigma}_g$ in high dimensions is infeasible, i.e., the number of free covariance parameters is $\mathcal{O}(p^2)$. To overcome this issue, one could assume a latent Gaussian factor structure for $\mathbf{X}$ within each mixture component — this is closely related to the factor regression model (FRM) of Y on $\mathbf{X}$ (cf. West, 2003; Wang et al., 2007;

Carvalho et al., 2008). Subedi et al. (2013) introduce the linear Gaussian cluster-weighted factor analyzers (CWFA) model, which has density

$$p(\mathbf{x}, y \mid \boldsymbol{\vartheta}) = \sum_{g=1}^{G} \pi_g \phi_1 \left(y \mid \beta_{0g} + \boldsymbol{\beta}'_{1g}\mathbf{x}, \sigma_g^2\right) \phi_p \left(\mathbf{x} \mid \boldsymbol{\mu}_g, \boldsymbol{\Lambda}_g \boldsymbol{\Lambda}'_g + \boldsymbol{\Psi}_g\right),$$

where $\boldsymbol{\Lambda}_g$ is a $p \times q$ matrix of factor loadings, with $q < p$, and $\boldsymbol{\Psi}_g$ is a $p \times p$ diagonal matrix with strictly positive diagonal entries. A family of 16 CWFA models follows by applying the PGMM covariance constraints in Table 3.1 as well as allowing the constraint $\sigma_g^2 = \sigma^2$. As was the case for the members of the PGMM family, the number of free covariance parameters in the CWFA models is linear in p. Note that Subedi et al. (2015) extend the CWFA models to the t-distribution.

9.7 Mixed-Type Data

9.7.1 A Mixture of Latent Variables Model

This monograph has largely focused on data that can be regarded as realizations of continuous random variables. However, other data types can be accommodated within the model-based clustering, classification, and discriminant analysis paradigms. To illustrate this point, consider a mixture of latent variables model for the model-based clustering and classification of data comprising variables of mixed type. This approach was proposed by Browne and McNicholas (2012) and can be regarded as a generalization of latent variable analysis (cf. Bartholomew and Knott, 1999). More specifically, it combines latent trait analysis and latent factor analysis into a single model and utilizes mixtures thereof. Note that Browne and McNicholas (2012) do not consider mixtures of latent class analyzers; however, they show that a mixture of latent class analyzers model can be written as a latent class analysis model.

In general, latent variable models assume that manifest $x_1, \ldots, x_p$ are independent given q-dimensional latent $\mathbf{y}$. Accordingly, the density function of the p-dimensional random variable $\mathbf{X} = (X_1, \ldots, X_p)'$ can be written

$$f(\mathbf{x}) = \int_{\mathcal{Y}} f(\mathbf{x} \mid \mathbf{y}, \boldsymbol{\theta}) h(\mathbf{y}) d\mathbf{y} = \int_{\mathcal{Y}} \prod_{j=1}^{p} g_i(x_j \mid \mathbf{y}, \boldsymbol{\theta}_j) h(\mathbf{y}) d\mathbf{y}, \qquad (9.10)$$

where $g(x_j \mid \mathbf{y}, \boldsymbol{\theta}_j)$ is the conditional distribution of X_j given $\mathbf{Y} = \mathbf{y}$ and parameters $\boldsymbol{\theta}_j$, and $h(\mathbf{y})$ is the marginal distribution of $\mathbf{Y}$. If X_j is categorical with levels $0, 1, \ldots, c_j - 1$, then the conditional distribution $X_j \mid \mathbf{y}, \boldsymbol{\theta}_j$ is Bernoulli with success probability given by

$$g(x_j = s \mid \mathbf{y}, \boldsymbol{\theta}_j) = \frac{\prod_{s=0}^{c_i - 1} \left[\exp\{\beta_{js0} + \boldsymbol{\beta}'_{js}\mathbf{y}\}\right]^{x_j(s)}}{1 + \sum_{k=1}^{c_j - 1} \exp\{\beta_{jk0} + \boldsymbol{\beta}'_{jk}\mathbf{y}\}},$$

where $x_j(s) = 1$ if the response falls into category s and $x_j(s) = 0$ otherwise, and $\boldsymbol{\beta}_{js} = (\beta_{js1}, \ldots, \beta_{jsq})'$ is a vector of coefficients. The category associated with $s = 0$ is known as the reference category. However, if X_j is continuous, then it is assumed to have a Gaussian conditional distribution with density

$$g(x_j \mid \mathbf{y}, \boldsymbol{\theta}_j) = \frac{1}{\sqrt{2\pi}\sigma_j} \exp\left\{ -\frac{1}{2\sigma_j^2} \left(x_j - \beta_{j0} - \boldsymbol{\beta}_j'\mathbf{y} \right)^2 \right\},$$

where $\boldsymbol{\beta}_j = (\beta_{j1}, \ldots, \beta_{jq})'$ is a vector of coefficients and $\beta_{j0} \in \mathbb{R}$.

For ease of notation, consider model-based clustering. The log-likelihood for p-dimensional $\mathbf{x}_1, \ldots, \mathbf{x}_n$ from the mixture of latent variables model can be written

$$
\begin{aligned}
l(\boldsymbol{\vartheta}) &= \sum_{i=1}^{n} \log \left\{ \int_{\mathcal{Y}} \left[\sum_{g=1}^{G} \pi_g f(\mathbf{x}_i \mid \mathbf{y}_i, \boldsymbol{\theta}_g) \right] h(\mathbf{y}_i) d\mathbf{y}_i \right\} \\
&= \sum_{i=1}^{n} \log \left\{ \sum_{g=1}^{G} \pi_g \left[\int_{\mathcal{Y}} \prod_{j=1}^{p} f(x_{ij} \mid \mathbf{y}_i, \boldsymbol{\theta}_{gj}) h(\mathbf{y}_i) d\mathbf{y}_i \right] \right\},
\end{aligned}
\tag{9.11}
$$

where, as usual, π_g is the gth mixing proportion and $\boldsymbol{\vartheta} = (\boldsymbol{\pi}, \boldsymbol{\theta}_1, \ldots, \boldsymbol{\theta}_G)$ denotes the model parameters. When the manifest variables have a categorical conditional distribution, the integral in (9.11) cannot be solved analytically; however, when the manifest variables have a Gaussian conditional distribution, it can be solved analytically. An EM algorithm is used for parameter estimation. Defining z_{ig} in the usual way, the complete-data comprise the observed $\mathbf{x}_1, \ldots, \mathbf{x}_n$ together with the latent $\mathbf{y}_{ig}$ and the component memberships z_{ig}, for $i = 1, \ldots, n$ and $g = 1, \ldots, G$. Then the complete-data log-likelihood is given by

$$l_c(\boldsymbol{\vartheta}) = \sum_{i=1}^{n} \sum_{g=1}^{G} z_{ig} \left[\log \pi_g + \sum_{j=1}^{p} \log f(x_{ij} \mid \mathbf{y}_{ig}, \boldsymbol{\theta}_{gj}) + \log h(\mathbf{y}_{ig}) \right]. \tag{9.12}$$

Details on the derivation of the parameter estimates are given by Browne and McNicholas (2012). In practice, models are fitted for a range of values of G and q, and the best model is selected using a criterion such as the BIC.

9.7.2   Illustration: Urinary System Disease Diagnosis

Czerniak and Zarzycki (2003) present data on two diseases of the urinary system: acute inflammation of the bladder and acute nephritis of renal pelvis origin. The data are designed to prepare an algorithm for an expert system that would then perform the presumptive diagnoses. They were created by a medical expert, and each case represents a potential patient. There are six predictor variables, one of which is interval (i.e., it can be regarded as a realization of a continuous random variable) and the remaining five are binary

Table 9.1

Six variables from the urinary disease diagnosis data, where all but "Temperature of patient" are binary (yes/no).

Temperature of patient (35–42°C)	Occurrence of nausea
Lumbar pain	Urine pushing
Burning of urethra, itch, swelling of urethra outlet	Micturition pains

(Table 9.1). The classes are inflammation and nephritis — either, neither, or both may be present.

Browne and McNicholas (2012) consider these data for clustering and classification. First, consider clustering. The mixture of latent variables model is fitted for $G = 1, \ldots, 5$ and $q = 1, 2, 3$. The BIC selects the model with $G = 2$ and $q = 1$, and the associated MAP classifications correspond perfectly to the inflammation diagnosis (Table 9.2).

Table 9.2

Cross-tabulation of respective diagnoses versus MAP classifications (A, B) from model-based clustering using the selected mixture of latent variables model.

		A	**B**
Inflammation	Yes	50	0
	No	0	70
Nephritis	Yes	19	40
	No	31	30

While this latent structure is clearly based on inflammation, a model-based classification approach can be used where the class of interest is inflammation, nephritis, or the combined diagnosis. Before proceeding to consider model-based classification analyses of these data, it is interesting to first observe the model-based clustering results in terms of the combined class (Table 9.3).

Browne and McNicholas (2012) carry out three model-based classification analyses. In each case, 100 runs are used where 25% of the instances are randomly treated as unlabelled on each run. In each case, preliminary runs clearly suggest that $q = 1$ latent variable should be used. When nephritis is taken to be the class, model-based classification using the mixture of latent variables model gives perfect classification on all 100 runs. The same is true when inflammation is taken to be the class. For the combined diagnosis class (inflammation, nephritis), there are now $G = 4$ components. The aggregate classification results for all 100 runs reflect an overall excellent classification performance (ARI = 0.957; Table 9.4).

The performance of the mixture of latent variables model for model-based

Table 9.3
Cross-tabulation of the combined diagnosis (inflammation, nephritis) versus
MAP classifications (A, B) from model-based clustering using the selected
mixture of latent variables model.

		A	B
	(Yes, Yes)	19	0
	(Yes, No)	31	0
(Inflammation, Nephritis)	(No, Yes)	0	40
	(No, No)	0	30

Table 9.4
Cross-tabulation of the combined diagnosis (inflammation, nephritis) versus
aggregate classifications (A–D) for the unlabelled instances from 100 model-
based classification runs using the mixture of latent variables model with
$G = 2$, $q = 1$, and 25% of instances treated as unlabelled at each run.

		A	B	C	D
	(Yes, Yes)	490	3	0	0
	(Yes, No)	37	711	0	0
(Inflammation, Nephritis)	(No, Yes)	5	0	1045	0
	(No, No)	16	0	0	693

classification illustrates the flexibility of the approach. With 25% of the instances unlabelled, the inflammation and nephritis classes, respectively, are
perfectly classified on all 100 runs. Furthermore, the combined class is picked
up with great accuracy, and only $q = 1$ latent variable is needed in each case.

9.8   Alternatives to the EM Algorithm

9.8.1   Variational Bayes Approximations

Variational Bayes approximations represent iterative Bayesian alternatives to
the EM algorithm. The fast and deterministic nature of the variational Bayes
approach has made it increasingly popular over the past decade or two (e.g.,
Waterhouse et al., 1996; Jordan et al., 1999; Corduneanu and Bishop, 2001).
The tractability of the variational approach allows for simultaneous model
selection and parameter estimation, thus removing the need for a criterion to
select the number of components G and reducing the associated computational
overhead. The variational Bayes algorithm has been applied to Gaussian mix-

ture models (e.g., Teschendorff et al., 2005; McGrory and Titterington, 2007; Subedi and McNicholas, 2016) as well as non-Gaussian mixtures (e.g., Subedi and McNicholas, 2014).

For data $\mathbf{x}$, the joint conditional distribution of parameters $\boldsymbol{\theta}$ and missing data $\mathbf{z}$ is approximated by constructing a tight lower bound on the marginal likelihood using a computationally convenient approximating density $q(\boldsymbol{\theta}, \mathbf{z})$. The density $q(\boldsymbol{\theta}, \mathbf{z})$ is obtained by minimizing the Kullback–Leibler (KL) divergence between the true density $p(\boldsymbol{\theta}, \mathbf{z} \mid \mathbf{x})$ and the approximating density (Beal, 2003; McGrory and Titterington, 2007), where

$$\mathrm{KL}(q(\boldsymbol{\theta}, \mathbf{z}) \mid p(\boldsymbol{\theta}, \mathbf{z} \mid \mathbf{x})) = \int_{\Theta} \sum_{\mathbf{z}} q(\boldsymbol{\theta}, \mathbf{z}) \log \left\{ \frac{q(\boldsymbol{\theta}, \mathbf{z})}{p(\boldsymbol{\theta}, \mathbf{z} \mid \mathbf{x})} \right\} d\boldsymbol{\theta}.$$

For computational convenience, the approximating density is restricted to have a factorized form so that

$$q(\boldsymbol{\theta}, \mathbf{z}) = q_{\boldsymbol{\theta}}(\boldsymbol{\theta}) q_{\mathbf{z}}(\mathbf{z}).$$

After choosing a conjugate prior, the appropriate approximating density $q_{\boldsymbol{\theta}}(\boldsymbol{\theta})$ can be obtained by solving a set of coupled non-linear equations. Because the KL divergence is non-negative, minimizing the KL divergence is equivalent to maximizing the lower bound.

When applied to mixture models, the algorithm is initialized with more components than are expected to be necessary. Then estimation of the parameters and the number of components is performed simultaneously. However, when applied within a family setting (e.g., see Subedi and McNicholas, 2016, in the case of the GPCM family), the member of the family must still be selected. Following the suggestion of McGrory and Titterington (2007), Subedi and McNicholas (2016) use the deviance information criterion (DIC; Spiegelhalter et al., 2002),

$$\mathrm{DIC} = -2 \log p(\mathbf{x} \mid \tilde{\boldsymbol{\theta}}) + 2p_D,$$

where

$$2p_D \approx -2 \int_{\Theta} q(\boldsymbol{\theta}) \log \left\{ \frac{q(\boldsymbol{\theta})}{p(\boldsymbol{\theta})} \right\} d\boldsymbol{\theta} + 2 \log \left\{ \frac{q(\tilde{\boldsymbol{\theta}})}{p(\tilde{\boldsymbol{\theta}})} \right\}$$

and $\log p(\mathbf{x} \mid \tilde{\boldsymbol{\theta}})$ is the posterior log-likelihood of the data.

9.8.2 Other Approaches

Bayesian approaches to mixture modelling offer the flexibility of sampling from computationally complex models using various MCMC sampling algorithms. While this has led to some interest in their use (Diebolt and Robert, 1994; Richardson and Green, 1997; Stephens, 1997, 2000; Casella et al., 2002), difficulties remain. Two such difficulties are computational overhead and convergence (Celeux et al., 2000; Jasra et al., 2005). In some respects, variational

Bayes approximations present an alternative to MCMC algorithms for mixture modelling parameter estimation by overcoming the hurdles of MCMC sampling while still retaining benefits of the Bayesian approach.

For some of the GPCM models, an alternative to the EM algorithm is required. It is most straightforward to consider the more general problem of minimization of the objective function

$$f(\mathbf{D}) = \sum_{g=1}^{G} \operatorname{tr}\{\mathbf{W}_g \mathbf{D} \mathbf{A}_g^{-1} \mathbf{D}'\}, \tag{9.13}$$

where $\mathbf{D}$ is an orthonormal matrix, e.g., an eigenvector matrix, $\mathbf{W}_1, \ldots, \mathbf{W}_G$ are positive-definite, e.g., sample covariance matrices, and $\mathbf{A}_1, \ldots, \mathbf{A}_G$ are diagonal matrices with positive diagonal elements. When minimizing $f(\mathbf{D})$, the search space is the orthogonal Stiefel manifold, which is an embedded submanifold of $\mathbb{R}^{p \times p}$ given by the set of all orthonormal matrices. Celeux and Govaert (1995) give an EM algorithm where each M-step performs the minimization in (9.13); the Flury method (Flury and Gautschi, 1986) is the most popular algorithm to minimize (9.13). The Flury method is well-known to be slow in all but low dimensions (e.g, Lefkomtch, 2004; Boik, 2007), which has obvious knock-on effects for the methods that draw upon it. To overcome this problem in the case of some of the GPCM models, Browne and McNicholas (2014c) implement an accelerated line search (ALS) for optimization on the orthogonal Stiefel manifold. This ALS outperforms the Flury method in high dimensions while also reducing the number of degenerate solutions produced by the EM algorithm; however, it does not exploit the convexity of the objective function. Browne and McNicholas (2014a) exploit this convexity property to obtain several simple majorization-minimization (MM) algorithms (cf. Hunter and Lange, 2000a; Hunter, 2004) following methodology from Kiers (2002). Parameter estimation for two of the GPCMs — EVE and VVE — is carried out using these MM algorithms within the `mixture` package.

9.9   Challenges and Open Questions

As a field of endeavour, mixture model-based approaches to clustering, classification, and discriminant analysis has made tremendous strides forward in the past decade or so. However, some important challenges and open questions remain. For one, the matter of model selection is still not resolved to a satisfactory extent. Although it is much maligned, the BIC remains the model selection criterion of choice. This point is reinforced by many applications within the literature as well as some dedicated comparative studies (e.g., Steele and Raftery, 2010). The averaging approach discussed in Section 9.3 provides an alternative to the "single best model" paradigm; however, it too depends on the BIC. Other alternatives include those based on hypothesis testing, e.g., Punzo et al. (2016). As the field moves away from Gaussian mixture models, questions about the efficacy of the BIC for mixture model selection will only grow in their frequency and intensity. The search for a more effective criterion, perhaps dedicated to clustering and classification applications, is perhaps the single greatest challenge within the field.

As is the case for model selection, the issue of the selection of starting values is not a new problem but it is persistent, and efforts in this direction are sure to continue. It is quite likely that the increasing ease of access to high-performance computing equipment will help dictate the direction this work takes. The increasing dimensionality and complexity of modern datasets raise issues that demand answers. For instance, there has been a paucity of work on clustering mixed type data (cf. Section 9.7), ordinal data (cf. Section 9.5), and binary data — a recent example of work on binary data is given by Tang et al. (2015), who consider clustering high-dimensional binary data (cf. Section 9.5). Another example is clickstream, and similar, data for which there has also been relatively little work (e.g., Melnykov, 2016).

The pursuit of more flexible models is sure to continue and has the potential to provide more useful tools; however, it is very important that such methods are accompanied by effective software. This reflects a general problem: there are far more promising methods for model-based clustering, classification, and discriminant analysis than there are effective software packages. Beyond what is mentioned in Chapter 8, only minimal work has been done on model-based approaches to longitudinal data (e.g., De la Cruz–Mesía et al., 2008, use a mixture of non-linear hierarchical models) and this area also merits further investigation. Some recent work on fractionally-supervised classification (Vrbik and McNicholas, 2015) is sure to spawn further work in similar directions. The use of copulas in mixture model-based approaches has already received some attention (e.g., Jajuga and Papla, 2006; Vrac et al., 2012; Kosmidis and Karlis, 2015; Marbac et al., 2015) and this sure to continue. Finally, there are some specific data types — both recently emerged and yet to emerge — that deserve their own special attention. One such type is next-generation sequencing data, which has already driven some interesting work within the field (e.g., Rau et al., 2015) and will surely continue to do so for some time.

Appendix

A.1 Linear Algebra Results

The following theorems are taken from Graybill (1983).

Theorem 1 *Let $\mathbf{A}$ and $\mathbf{B}$ be any matrices such that $\mathbf{AB}$ is defined; then*

$$(\mathbf{AB})' = \mathbf{B}'\mathbf{A}'.$$

Theorem 2 *If $\mathbf{A}$ is any matrix, then $\mathbf{A}'\mathbf{A}$ and $\mathbf{AA}'$ are symmetric.*

Theorem 3 *If $\mathbf{A}$ is a nonsingular matrix, then $\mathbf{A}'$ and $\mathbf{A}^{-1}$ are nonsingular and*

$$\left(\mathbf{A}'\right)^{-1} = \left(\mathbf{A}^{-1}\right)'.$$

Theorem 4 *If each element of the ith row of an $n \times n$ matrix $\mathbf{A}$ contains a given factor k, then we may write $|\mathbf{A}| = k|\mathbf{B}|$, where the rows of $\mathbf{B}$ are the same as the rows of $\mathbf{A}$ except that the number k has been factored from each element of the ith row of $\mathbf{A}$.*

A corollary of Theorem 4, given below as Corollary 1, it is also stated as a result by Anton and Rorres (1994).

Corollary 1 *Let $\mathbf{A}$ and $\mathbf{B}$ be $n \times n$ matrices such that*

$$\mathbf{A} = k\mathbf{B},$$

where k is a scalar. Then

$$|\mathbf{A}| = k^n|\mathbf{B}|.$$

From this corollary, it follows that, for $a \in \mathbb{R}^+$ and an $n \times n$ identity matrix $\mathbf{I}_n$,

$$\log|a^{-1}\mathbf{I}_n| = \log a^{-n} + \log|\mathbf{I}_n| = n \log a^{-1}.$$

The following results are taken from Lütkepohl (1996, Chapter 4) and are also available elsewhere:

$$\mathbf{A}_{m \times n}, \mathbf{B}_{n \times m} : \ \mathrm{tr}\{\mathbf{AB}\} = \mathrm{tr}\{\mathbf{BA}\}.$$
$$\mathbf{A}_{m \times m}, \mathbf{B}_{m \times m} : |\ \mathbf{AB}\ | = |\ \mathbf{A}\ ||\ \mathbf{B}\ |.$$
$$\mathbf{A}_{m \times m} : |\ \mathbf{A}'\ | = |\ \mathbf{A}\ |.$$
$$\mathbf{A}_{m \times m}, \text{ nonsingular} : |\ \mathbf{A}^{-1}\ | = |\ \mathbf{A}\ |^{-1}.$$
$$\mathbf{A}_{m \times m} = [a_{ij}], \text{ triangular} : |\ \mathbf{A}\ | = \prod_{i=1}^{m} a_{ii}.$$

A.2 Matrix Calculus Results

We assume that all matrices and vectors are real, all objects that are differentiated are continuously differentiable, and all differentials are well defined. The following results, taken from Lütkepohl (1996, Chapter 10) and also avaiable elsewhere, are utilized herein:

$$\mathbf{X}_{m \times m} \text{ nonsingular}: \quad \frac{\partial \log |\mathbf{X}|}{\partial \mathbf{X}} = (\mathbf{X}')^{-1}.$$

$$\mathbf{X}_{m \times n}, \mathbf{A}_{n \times m}: \quad \frac{\partial \operatorname{tr}\{\mathbf{X}\mathbf{A}\}}{\partial \mathbf{X}} = \frac{\partial \operatorname{tr}\{\mathbf{A}\mathbf{X}\}}{\partial \mathbf{X}} = \mathbf{A}'.$$

$$\mathbf{X}_{m \times n}, \mathbf{A}_{m \times n}: \quad \frac{\partial \operatorname{tr}\{\mathbf{X}'\mathbf{A}\}}{\partial \mathbf{X}} = \frac{\partial \operatorname{tr}\{\mathbf{A}\mathbf{X}'\}}{\partial \mathbf{X}} = \mathbf{A}.$$

$$\mathbf{X}_{m \times n}, \mathbf{A}_{p \times m}, \mathbf{B}_{n \times p}: \quad \frac{\partial \operatorname{tr}\{\mathbf{A}\mathbf{X}\mathbf{B}\}}{\partial \mathbf{X}} = \mathbf{A}'\mathbf{B}'.$$

$$\mathbf{X}_{m \times n}, \mathbf{A}_{n \times n} \text{ symmetric}: \quad \frac{\partial \operatorname{tr}\{\mathbf{X}\mathbf{A}\mathbf{X}'\}}{\partial \mathbf{X}} = 2\mathbf{X}\mathbf{A}.$$

$$\mathbf{X}_{m \times n}, \mathbf{A}_{n \times m}, \mathbf{B}_{n \times m}: \quad \frac{\partial \operatorname{tr}\{\mathbf{X}\mathbf{A}\mathbf{X}\mathbf{B}\}}{\partial \mathbf{X}} = \mathbf{B}'\mathbf{X}'\mathbf{A}' + \mathbf{A}'\mathbf{X}'\mathbf{B}'.$$

$$\mathbf{X}_{m \times n}, \mathbf{A}_{n \times n}, \mathbf{B}_{m \times m}: \quad \frac{\partial \operatorname{tr}\{\mathbf{X}\mathbf{A}\mathbf{X}'\mathbf{B}\}}{\partial \mathbf{X}} = \mathbf{B}'\mathbf{X}\mathbf{A}' + \mathbf{B}\mathbf{X}\mathbf{A}.$$

$$\mathbf{X}_{m \times n}, \mathbf{A}_{p \times m}, \mathbf{B}_{m \times p}: \quad \frac{\partial \operatorname{tr}\{\mathbf{A}\mathbf{X}\mathbf{X}'\mathbf{B}\}}{\partial \mathbf{X}} = (\mathbf{B}\mathbf{A} + \mathbf{A}'\mathbf{B}')\mathbf{X}.$$

$$\mathbf{X}_{n \times n} \text{ nonsingular}: \quad \frac{\partial |\mathbf{X}^{-1}|}{\partial \mathbf{X}} = -|\mathbf{X}|^{-1}(\mathbf{X}')^{-1}.$$

A related result is also useful, i.e.,

$$\mathbf{X}_{n \times n} \text{ nonsingular}: \quad \frac{\partial |\mathbf{X}|}{\partial \mathbf{X}^{-1}} = -|\mathbf{X}|\mathbf{X}'.$$

Finally, it might be of use to consider two more general results that encapsulate, as special cases, some of the aforementioned results:

$$\mathbf{X}_{m \times n}, \mathbf{A}_{p \times m}, \mathbf{B}_{n \times m}, \mathbf{C}_{n \times p}: \quad \frac{\partial \operatorname{tr}\{\mathbf{A}\mathbf{X}\mathbf{B}\mathbf{X}\mathbf{C}\}}{\partial \mathbf{X}} = \mathbf{A}'\mathbf{C}'\mathbf{X}'\mathbf{B}' + \mathbf{B}'\mathbf{X}'\mathbf{A}'\mathbf{C}'.$$

$$\mathbf{X}_{m \times n}, \mathbf{A}_{p \times m}, \mathbf{B}_{n \times n}, \mathbf{C}_{m \times p}: \quad \frac{\partial \operatorname{tr}\{\mathbf{A}\mathbf{X}\mathbf{B}\mathbf{X}'\mathbf{C}\}}{\partial \mathbf{X}} = \mathbf{A}'\mathbf{C}'\mathbf{X}\mathbf{B}' + \mathbf{C}\mathbf{A}\mathbf{X}\mathbf{B}.$$

Again, these results are taken from Lütkepohl (1996, Chapter 10).

A.3   Method of Lagrange Multipliers

Lagrange (1788) introduced a method of finding local extrema of a function subject to some constraint, which is also expressed as a function. This method is summarized here, and described in detail by Fraleigh (1990).

To find local extrema of a function $f(x, y, z)$, subject to the constraint $g(x, y, z) = 0$, note that there exists some constant ι such that

$$\nabla f = \iota(\nabla g),$$

where ι is called the Lagrange multiplier. This, together with the constraint $g(x, y, z) = 0$, leads to what Fraleigh (1990) refers to as the "conditions of the method of Lagrange multipliers":

$$\frac{\partial f}{\partial x} = \iota \frac{\partial g}{\partial x}, \quad \frac{\partial f}{\partial y} = \iota \frac{\partial g}{\partial y}, \quad \frac{\partial f}{\partial z} = \iota \frac{\partial g}{\partial z}, \quad \text{and} \quad g(x, y, z) = 0.$$

Furthermore, writing

$$L(x, y, z, \iota) = f(x, y, z) - \iota g(x, y, z)$$

makes these conditions equivalent to

$$\frac{\partial L}{\partial x} = 0, \quad \frac{\partial L}{\partial y} = 0, \quad \frac{\partial L}{\partial z} = 0, \quad \text{and} \quad \frac{\partial L}{\partial \iota} = 0.$$

References

Absil, P.–A., R. Mahony, and R. Sepulchre (2008). *Optimization Algorithms on Matrix Manifolds*. Princeton: Princeton University Press.

Aitken, A. C. (1926). A series formula for the roots of algebraic and transcendental equations. *Proceedings of the Royal Society of Edinburgh 45*(1), 14–22.

Aitkin, M., D. Anderson, and J. Hinde (1981). Statistical modelling of data on teaching styles. *Journal of the Royal Statistical Society: Series A 144*(4), 419–461.

Aitkin, M. and G. T. Wilson (1980). Mixture models, outliers, and the EM algorithm. *Technometrics 22*(3), 325–331.

Ali, M. M., J. Woo, and S. Nadarajah (2010). Some skew symmetric inverse reflected distributions. *Brazilian Journal of Probability and Statistics 24*(1), 1–23.

Alon, U., N. Barkai, D. A. Notterman, K. Gish, S. Ybarra, D. Mack, and A. J. Levine (1999). Broad patterns of gene expression revealed by clustering analysis of tumor and normal colon tissues probed by oligonucleotide arrays. *Proceedings of the National Academy of Sciences of the United States of America 96*(12), 6745–6750.

Altman, E. I. (1968). Financial ratios, discriminant analysis and the prediction of corporate bankruptcy. *Journal of Finance 23*(4), 589–609.

Anderlucci, L. and C. Viroli (2015). Covariance pattern mixture models for multivariate longitudinal data. *The Annals of Applied Statistics 9*(2), 777–800.

Andrews, J. L. and P. D. McNicholas (2011a). Extending mixtures of multivariate t-factor analyzers. *Statistics and Computing 21*(3), 361–373.

Andrews, J. L. and P. D. McNicholas (2011b). Mixtures of modified t-factor analyzers for model-based clustering, classification, and discriminant analysis. *Journal of Statistical Planning and Inference 141*(4), 1479–1486.

Andrews, J. L. and P. D. McNicholas (2012). Model-based clustering, classification, and discriminant analysis via mixtures of multivariate *t*-distributions: The *t*EIGEN family. *Statistics and Computing 22*(5), 1021–1029.

Andrews, J. L. and P. D. McNicholas (2013). *vscc: Variable Selection for Clustering and Classification*. R package version 0.2.

Andrews, J. L. and P. D. McNicholas (2014). Variable selection for clustering and classification. *Journal of Classification 31*(2), 136–153.

Andrews, J. L. and P. D. McNicholas (2015). *teigen: Model-Based Clustering and Classification with the Multivariate t Distribution*. R package version 2.0.8.

Andrews, J. L., P. D. McNicholas, and S. Subedi (2011). Model-based classification via mixtures of multivariate t-distributions. *Computational Statistics and Data Analysis 55*(1), 520–529.

Anton, H. and C. Rorres (1994). *Elementary Linear Algebra* (7th ed.). New York: John Wiley & Sons.

Arabie, P. and S. A. Boorman (1973). Multidimensional scaling of measures of distance between partitions. *Journal of Mathematical Psychology 10*(2), 148–203.

Azzalini, A., R. P. Browne, M. G. Genton, and P. D. McNicholas (2016). On nomenclature for, and the relative merits of, two formulations of skew distributions. *Statistics and Probability Letters 110*, 201–206.

Azzalini, A. and A. Capitanio (1999). Statistical applications of the multivariate skew normal distribution. *Journal of the Royal Statistical Society: Series B 61*(3), 579–602.

Azzalini, A. and A. Capitanio (2003). Distributions generated by perturbation of symmetry with emphasis on a multivariate skew *t* distribution. *Journal of the Royal Statistical Society: Series B 65*(2), 367–389.

Azzalini, A. with the collaboration of A. Capitanio (2014). *The Skew-Normal and Related Families*. IMS monographs. Cambridge: Cambridge University Press.

Azzalini, A. and A. D. Valle (1996). The multivariate skew-normal distribution. *Biometrika 83*, 715–726.

Baek, J. and G. J. McLachlan (2011). Mixtures of common t-factor analyzers for clustering high-dimensional microarray data. *Bioinformatics 27*, 1269–1276.

Baek, J., G. J. McLachlan, and L. K. Flack (2010). Mixtures of factor analyzers with common factor loadings: Applications to the clustering and visualization of high-dimensional data. *IEEE Transactions on Pattern Analysis and Machine Intelligence 32*, 1298–1309.

Banfield, J. D. and A. E. Raftery (1993). Model-based Gaussian and non-Gaussian clustering. *Biometrics 49*(3), 803–821.

Baricz, A. (2010). Turn type inequalities for some probability density functions. *Studia Scientiarum Mathematicarum Hungarica 47*, 175–189.

Barndorff–Nielsen, O. (1978). Hyperbolic distributions and distributions on hyperbolae. *Scandinavian Journal of Statistics 5*(3), 151–157.

Barndorff–Nielsen, O. and C. Halgreen (1977). Infinite divisibility of the hyperbolic and generalized inverse Gaussian distributions. *Zeitschrift für Wahrscheinlichkeitstheorie und Verwandte Gebiete 38*, 309–311.

Barndorff–Nielsen, O., J. Kent, and M. Sørensen (1982). Normal variance-mean mixtures and z distributions. *International Statistical Review 50*(2), 145–159.

Barnett, V. (1999). *Comparative Statistical Inference* (3rd ed.). Chichester: Wiley.

Bartholomew, D. and M. Knott (1999). *Latent Variable Models and Factor Analysis* (2nd ed.). Kendall's Library of Statistics. London: Arnold.

Bartlett, M. S. (1953). Factor analysis in psychology as a statistician sees it. In *Uppsala Symposium on Psychological Factor Analysis*, Number 3 in Nordisk Psykologi's Monograph Series, pp. 23–34. Copenhagen: Ejnar Mundsgaards.

Baudry, J.–P., A. E. Raftery, G. Celeux, K. Lo, and R. Gottardo (2010). Combining mixture components for clustering. *Journal of Computational and Graphical Statistics 19*(2), 332–353.

Baum, L. E., T. Petrie, G. Soules, and N. Weiss (1970). A maximization technique occurring in the statistical analysis of probabilistic functions of Markov chains. *Annals of Mathematical Statistics 41*, 164–171.

Beal, M. J. (2003). *Variational Algorithms for Approximate Bayesian Inference*. Ph.D. thesis, University of London.

Benoît (1924). Note sur une méthode de résolution des équations normales provenant de l'application de la méthode des moindres carrés à un système d'équations linéaires en nombre inférieur celui des inconnues (Procédé du Commandant Cholesky). *Bulletin Géodésique 2*, 67–77.

Bergé, L., C. Bouveyron, and S. Girard (2012). HDclassif: An R package for model-based clustering and discriminant analysis of high-dimensional data. *Journal of Statistical Software 46*(6), 1–29.

Bergé, L., C. Bouveyron, and S. Girard (2013). *HDclassif: High Dimensional Discriminant Analysis and Data Clustering.* R package version 1.2.3.

Bhattacharya, S. and P. D. McNicholas (2014). A LASSO-penalized BIC for mixture model selection. *Advances in Data Analysis and Classification 8*(1), 45–61.

Biernacki, C., G. Celeux, and G. Govaert (2000). Assessing a mixture model for clustering with the integrated completed likelihood. *IEEE Transactions on Pattern Analysis and Machine Intelligence 22*(7), 719–725.

Biernacki, C., G. Celeux, and G. Govaert (2003). Choosing starting values for the EM algorithm for getting the highest likelihood in multivariate Gaussian mixture models. *Computational Statistics and Data Analysis 41*, 561–575.

Biernacki, C., G. Celeux, G. Govaert, and F. Langrognet (2006). Model-based cluster and discriminant analysis with the MIXMOD software. *Computational Statistics and Data Analysis 51*(2), 587–600.

Biernacki, C., G. Celeux, and G. Govaert (2010). Exact and Monte Carlo calculations of integrated likelihoods for the latent class model. *Journal of Statistical Planning and Inference 140*(11), 2991–3002.

Blæsild, P. (1978). The shape of the generalized inverse Gaussian and hyperbolic distributions. Research Report 37, Department of Theoretical Statistics, Aarhus University, Denmark.

Böhning, D., E. Dietz, R. Schaub, P. Schlattmann, and B. Lindsay (1994). The distribution of the likelihood ratio for mixtures of densities from the one-parameter exponential family. *Annals of the Institute of Statistical Mathematics 46*, 373–388.

Boik, R. J. (2007). Spectral models for covariance matrices. *Biometrika 89*, 159–182.

Bombrun, L., F. Pascal, J.–Y. Tourneret, and Y. Berthoumieu (2012). Performance of the maximum likelihood estimators for the parameters of multivariate generalized Gaussian distributions. In *2012 IEEE International Conference on Acoustics, Speech and Signal Processing (ICASSP)*, pp. 3525–3528. IEEE.

Bouveyron, C. and C. Brunet–Saumard (2014). Model-based clustering of high-dimensional data: A review. *Computational Statistics and Data Analysis 71*, 52–78.

Bouveyron, C., G. Celeux, and S. Girard (2011). Intrinsic dimension estimation by maximum likelihood in isotropic probabilistic PCA. *Pattern Recognition Letters 32*(14), 1706–1713.

Bouveyron, C., S. Girard, and C. Schmid (2007a). High-dimensional data clustering. *Computational Statistics and Data Analysis 52*(1), 502–519.

Bouveyron, C., S. Girard, and C. Schmid (2007b). High dimensional discriminant analysis. *Communications in Statistics – Theory and Methods 36*(14), 2607–2623.

Branco, M. D. and D. K. Dey (2001). A general class of multivariate skew-elliptical distributions. *Journal of Multivariate Analysis 79*, 99–113.

Browne, R. P. and P. D. McNicholas (2012). Model-based clustering and classification of data with mixed type. *Journal of Statistical Planning and Inference 142*(11), 2976–2984.

Browne, R. P. and P. D. McNicholas (2014a). Estimating common principal components in high dimensions. *Advances in Data Analysis and Classification 8*(2), 217–226.

Browne, R. P. and P. D. McNicholas (2014b). *mixture: Mixture Models for Clustering and Classification*. R package version 1.1.

Browne, R. P. and P. D. McNicholas (2014c). Orthogonal Stiefel manifold optimization for eigen-decomposed covariance parameter estimation in mixture models. *Statistics and Computing 24*(2), 203–210.

Browne, R. P. and P. D. McNicholas (2015). A mixture of generalized hyperbolic distributions. *Canadian Journal of Statistics 43*(2), 176–198.

Browne, R. P., P. D. McNicholas, and M. D. Sparling (2012). Model-based learning using a mixture of mixtures of Gaussian and uniform distributions. *IEEE Transactions on Pattern Analysis and Machine Intelligence 34*(4), 814–817.

Cagnone, S. and C. Viroli (2012). A factor mixture analysis model for multivariate binary data. *Statistical Modelling 12*(3), 257–277.

Campbell, J. G., C. Fraley, F. Murtagh, and A. E. Raftery (1997). Linear flaw detection in woven textiles using model-based clustering. *Pattern Recognition Letters 18*, 1539–1548.

Campbell, N. A. (1984). Mixture models and atypical values. *Mathematical Geology 16*(5), 465–477.

Carvalho, C., J. Chang, J. Lucas, J. Nevins, Q. Wang, and M. West (2008). High-dimensional sparse factor modeling: Applications in gene expression genomics. *Journal of the American Statistical Association 103*(484), 1438–1456.

Casella, G., K. L. Mengersen, C. P. Robert, and D. M. Titterington (2002). Perfect samplers for mixtures of distributions. *Journal of the Royal Statistical Society: Series B 64*, 777–790.

Cattell, R. B. (1957). *Personality and Motivation Structure and Measurement*. New York: World Book.

Celeux, G. and G. Govaert (1991). Clustering criteria for discrete data and latent class models. *Journal of Classification 8*(2), 157–176.

Celeux, G. and G. Govaert (1992). A classification EM algorithm for clustering and two stochastic versions. *Computational Statistics and Data Analysis 14*(3), 315–332.

Celeux, G. and G. Govaert (1995). Gaussian parsimonious clustering models. *Pattern Recognition 28*(5), 781–793.

Celeux, G., M. Hurn, and C. P. Robert (2000). Computational and inferential difficulties with mixture posterior distributions. *Journal of the American Statistical Association 95*, 957–970.

Celeux, G., M.–L. Martin–Magniette, C. Maugis–Rabusseau, and A. E. Raftery (2014). Comparing model selection and regularization approaches to variable selection in model-based clustering. *Journal de la Société Française de Statistique 155*(2), 57–71.

Chen, X., S. T. Cheung, S. So, S. T. Fan, C. Barry, J. Higgins, K.-M. Lai, J. Ji, S. Dudoit, I. O. Ng, M. van de Rijn, D. Botstein, and P. O. Brown (2002). Gene expression patterns in human liver cancers. *Molecular Biology of the Cell 13*(6), 1929–1939.

Cho, D. and T. D. Bui (2005). Multivariate statistical modeling for image denoising using wavelet transforms. *Signal Processing: Image Communication 20*(1), 77–89.

Cook, D. and D. F. Swayne (2007). *Interactive and Dynamic Graphics for Data Analysis: With R and GGobi*. New York: Springer.

Corduneanu, A. and C. M. Bishop (2001). Variational Bayesian model selection for mixture distributions. In *Artificial Intelligence and Statistics*, pp. 27–34. Los Altos, CA: Morgan Kaufmann.

Cormack, R. M. (1971). A review of classification (with discussion). *Journal of the Royal Statistical Society: Series A 34*, 321–367.

Czerniak, J. and H. Zarzycki (2003). Application of rough sets in the presumptive diagnosis of urinary system diseases. In *Artificial Intelligence and Security in Computing Systems, ACS 2002 9th International Conference Proceedings*, Norwell, MA, pp. 41–53. Kluwer Academic Publishers.

Dang, U. J., R. P. Browne, and P. D. McNicholas (2015). Mixtures of multivariate power exponential distributions. *Biometrics 71*(4), 1081–1089.

Dasgupta, A. and A. E. Raftery (1998). Detecting features in spatial point processes with clutter via model-based clustering. *Journal of the American Statistical Association 93*, 294–302.

Day, N. E. (1969). Estimating the components of a mixture of normal distributions. *Biometrika 56*, 463–474.

De la Cruz–Mesía, R., F. A. Quintana, and G. Marshall (2008). Model-based clustering for longitudinal data. *Computational Statistics and Data Analysis 52*(3), 1441–1457.

De Leeuw, J. (1994). Information systems and data analysis. In H. H. Bock, W. Lenski, and M. M. Richter (Eds.), *Block Relaxation Algorithms in Statistics*, pp. 308–325. Berlin: Springer-Verlag.

de Leon, A. R. (2005). Pairwise likelihood approach to grouped continuous model and its extension. *Statistics and Probability Letters 75*(1), 49–57.

De Veaux, R. D. and A. M. Krieger (1990). Robust estimation of a normal mixture. *Statistics and Probability Letters 10*(1), 1–7.

Dean, N., A. E. Raftery, and L. Scrucca (2012). *clustvarsel: Variable Selection for Model-Based Clustering*. R package version 2.0.

Demarta, S. and A. J. McNeil (2005). The t copula and related copulas. *International Statistical Review 73*(1), 111–129.

Dempster, A. P., N. M. Laird, and D. B. Rubin (1977). Maximum likelihood from incomplete data via the EM algorithm. *Journal of the Royal Statistical Society: Series B 39*(1), 1–38.

Diebolt, J. and C. Robert (1994). Estimation of finite mixture distributions through Bayesian sampling. *Journal of the Royal Statistical Society: Series B 56*, 363–375.

Downton, M. and T. Brennan (1980, June). Comparing classifications: An evaluation of several coefficients of partition agreement. Paper presented at the meeting of the Classification Society, Boulder, CO.

Edwards, A. W. F. and L. L. Cavalli–Sforza (1965). A method for cluster analysis. *Biometrics 21*, 362–375.

Everitt, B. S. and D. J. Hand (1981). *Finite Mixture Distributions*. Monographs on Applied Probability and Statistics. London: Chapman and Hall.

Everitt, B. S., S. Landau, M. Leese, and D. Stahl (2011). *Cluster Analysis* (5th ed.). Chichester: John Wiley & Sons.

Fabrigar, L. R., D. T. Wegener, R. C. MacCallum, and E. J. Strahan (1999). Evaluating the use of exploratory factor analysis in psychological research. *Psychological Methods 4*(3), 272–299.

Fan, J. and R. Li (2001). Variable selection via nonconcave penalized likelihood and its oracle properties. *Journal of the American Statistical Association 96*(456), 1348–1360.

Flury, B. (1988). *Common Principal Components and Related Multivariate Models*. New York: Wiley.

Flury, B. (2012). *Flury: Data Sets from Flury, 1997*. R package version 0.1-3.

Flury, B. W. and W. Gautschi (1986). An algorithm for simultaneous orthogonal transformation of several positive definite symmetric matrices to nearly diagonal form. *Journal on Scientific and Statistical Computing 7*(1), 169–184.

Forbes, F. and D. Wraith (2014). A new family of multivariate heavy-tailed distributions with variable marginal amounts of tailweights: Application to robust clustering. *Statistics and Computing 24*(6), 971–984.

Forina, M., C. Armanino, S. Lanteri, and E. Tiscornia (1983). Classification of olive oils from their fatty acid composition. In H. Martens and H. Russwurm Jr (Eds.), *Food Research and Data Analysis*, pp. 189–214. London: Applied Science Publishers.

Forina, M. and E. Tiscornia (1982). Pattern recognition methods in the prediction of Italian olive oil origin by their fatty acid content. *Annali di Chimica 72*, 143–155.

Fowlkes, E. B. and C. L. Mallows (1983). A method for comparing two hierarchical clusterings. *Journal of the American Statistical Association 78*, 553–569.

Fraleigh, J. B. (1990). *Calculus with Analytic Geometry* (3rd ed.). Reading, MA: Addison-Wesley.

Fraley, C. and A. E. Raftery (1998). How many clusters? Which clustering methods? Answers via model-based cluster analysis. *The Computer Journal 41*(8), 578–588.

Fraley, C. and A. E. Raftery (1999). MCLUST: Software for model-based cluster analysis. *Journal of Classification 16*, 297–306.

Fraley, C. and A. E. Raftery (2002a). MCLUST: Software for model-based clustering, density estimation, and discriminant analysis. Technical Report 415, University of Washington, Department of Statistics.

Fraley, C. and A. E. Raftery (2002b). Model-based clustering, discriminant analysis, and density estimation. *Journal of the American Statistical Association 97*(458), 611–631.

Fraley, C., A. E. Raftery, T. B. Murphy, and L. Scrucca (2012). mclust version 4 for R: Normal mixture modeling for model-based clustering, classification, and density estimation. Technical Report 597, Department of Statistics, University of Washington, Seattle, WA.

Fraley, C., A. E. Raftery, and L. Scrucca (2014). *mclust: Normal Mixture Modeling for Model-Based Clustering, Classification, and Density Estimation.* R package version 4.3.

Franczak, B. C., R. P. Browne, and P. D. McNicholas (2014). Mixtures of shifted asymmetric Laplace distributions. *IEEE Transactions on Pattern Analysis and Machine Intelligence 36*(6), 1149–1157.

Franczak, B. C., C. Tortora, R. P. Browne, and P. D. McNicholas (2015). Unsupervised learning via mixtures of skewed distributions with hypercube contours. *Pattern Recognition Letters 58*(1), 69–76.

Friedman, H. P. and J. Rubin (1967). On some invariant criteria for grouping data. *Journal of the American Statistical Association 62*, 1159–1178.

Fritz, H., L. A. García–Escudero, and A. Mayo–Iscar (2012). tclust: An R package for a trimming approach to cluster analysis. *Journal of Statistical Software 47*(12), 1–26.

Frühwirth–Schnatter, S. (2006). *Finite Mixture and Markov Switching Models.* New York: Springer-Verlag.

García–Escudero, L. A., A. Gordaliza, C. Matrn, and A. Mayo–Iscar (2008). A general trimming approach to robust cluster analysis. *The Annals of Statistics 36*(3), 1324–1345.

Gershenfeld, N. (1997). Nonlinear inference and cluster-weighted modeling. *Annals of the New York Academy of Sciences 808*(1), 18–24.

Getz, G., E. Levine, and E. Domany (2000). Coupled two-way clustering analysis of gene microarray data. *Proceedings of the National Academy of Sciences of the United States of America 97*(22), 12079–12084.

Ghahramani, Z. and G. E. Hinton (1997). The EM algorithm for factor analyzers. Technical Report CRG-TR-96-1, University of Toronto, Toronto, Canada.

Gneiting, T. (1997). Normal scale mixtures and dual probability densities. *Journal of Statistical Computation and Simulation 59*(4), 375–384.

Gollini, I. and T. B. Murphy (2014). Mixture of latent trait analyzers for model-based clustering of categorical data. *Statistics and Computing 24*(4), 569–588.

Golub, T. R., D. K. Slonim, P. Tamayo, C. Huard, M. Gaasenbeek, J. P. Mesirov, H. Coller, M. L. Loh, J. R. Downing, M. A. Caligiuri, C. D. Bloomfield, and E. S. Lander (1999). Molecular classification of cancer: Class discovery and class prediction by gene expression monitoring. *Science 286*, 531–537.

Gómez, E., M. A. Gomez–Viilegas, and J. M. Marin (1998). A multivariate generalization of the power exponential family of distributions. *Communications in Statistics – Theory and Methods 27*(3), 589–600.

Gómez–Sánchez–Manzano, E., M. A. Gómez–Villegas, and J. M. Marín (2008). Multivariate exponential power distributions as mixtures of normal distributions with Bayesian applications. *Communications in Statistics – Theory and Methods 37*(6), 972–985.

Goodman, L. (1974). Exploratory latent structure analysis using both identifiable and unidentifiable models. *Biometrika 61*(2), 215–231.

Gordon, A. D. (1981). *Classification*. London: Chapman and Hall.

Gower, J. C. (1971). Discussion of 'A review of classification' by Cormack. *Journal of the Royal Statistical Society: Series A 34*, 360–365.

Graybill, F. A. (1983). *Matrices with Applications in Statistics* (2nd ed.). Belmont, CA: Wadsworth.

Halgreen, C. (1979). Self-decomposibility of the generalized inverse Gaussian and hyperbolic distributions. *Zeitschrift für Wahrscheinlichkeitstheorie und Verwandte Gebiete 47*, 13–18.

Hartigan, J. A. (1975). *Clustering Algorithms*. New York: Wiley.

Hartigan, J. A. and M. A. Wong (1979). A k-means clustering algorithm. *Applied Statistics 28*(1), 100–108.

Hastie, T. and R. Tibshirani (1996). Discriminant analysis by Gaussian mixtures. *Journal of the Royal Statistical Society: Series B 58*(1), 155–176.

Hathaway, R. J. (1985). A constrained formulation of maximum likelihood estimation for normal mixture distributions. *The Annals of Statistics 13*(2), 795–800.

Heinz, G., L. J. Peterson, R. W. Johnson, and C. J. Kerk (2003). Exploring relationships in body dimensions. *Journal of Statistics Education 11*(2).

Heiser, W. J. (1995). Recent advances in descriptive multivariate analysis. In W. J. Krzanowski (Ed.), *Convergent Computation by Iterative Majorization: Theory and Applications in Multidimensional Data Analysis*, pp. 157–189. Oxford: Oxford University Press.

Hennig, C. (2000). Identifiablity of models for clusterwise linear regression. *Journal of Classification 17*(2), 273–296.

Hennig, C. (2004). Breakdown points for maximum likelihood estimators of location-scale mixtures. *The Annals of Statistics 32*(4), 1313–1340.

Hennig, C. (2010). Methods for merging Gaussian mixture components. *Advances in Data Analysis and Classification 4*, 3–34.

Hoeting, J. A., D. Madigan, A. E. Raftery, and C. T. Volinsky (1999). Bayesian model averaging: A tutorial. *Statistical Science 14*(4), 382–401.

Horn, J. L. (1965). A rationale and technique for estimating the number of factors in factor analysis. *Psychometrika 30*, 179–185.

Hu, W. (2005). *Calibration of Multivariate Generalized Hyperbolic Distributions Using the EM Algorithm, with Applications in Risk Management, Portfolio Optimization and Portfolio Credit Risk.* Ph.D. thesis, The Florida State University, Tallahassee.

Huber, P. J. (1964). Robust estimation of a location parameter. *The Annals of Mathematical Statistics 35*, 73–101.

Huber, P. J. (1981). *Robust Statistics.* New York: Wiley.

Hubert, L. and P. Arabie (1985). Comparing partitions. *Journal of Classification 2*(1), 193–218.

Hubert, L. J. (1977). Nominal scale response agreement as a generalized correlation. *British Journal of Mathematical and Statistical Psychology 30*, 98–103.

Humbert, S., S. Subedi, J. Cohn, B. Zeng, Y.-M. Bi, X. Chen, T. Zhu, P. D. McNicholas, and S. J. Rothstein (2013). Genome-wide expression profiling of maize in response to individual and combined water and nitrogen stresses. *BMC Genetics 14*(3).

Humphreys, L. G. and D. R. Ilgen (1969). Note on a criterion for the number of common factors. *Educational and Psychological Measurements 29*, 571–578.

Humphreys, L. G. and R. G. Montanelli, Jr. (1975). An investigation of the parallel analysis criterion for determining the number of common factors. *Multivariate Behavioral Research 10*, 193–205.

Hunter, D. R. and K. Lange (2000a). Quantile regression via an MM algorithm. *Journal of Computational and Graphical Statistics 9*, 60–77.

Hunter, D. R. and K. Lange (2000b). Rejoinder to discussion of "Optimization transfer using surrogate objective functions". *Journal of Computational and Graphical Statistics 9*, 52–59.

Hunter, D. R. (2004). MM algorithms for generalized Bradley-Terry models. *The Annals of Statistics 32*(1), 384–406.

Hurley, C. (2004). Clustering visualizations of multivariate data. *Journal of Computational and Graphical Statistics 13*(4), 788–806.

Ingrassia, S., S. C. Minotti, and A. Punzo (2014). Model-based clustering via linear cluster-weighted models. *Computational Statistics and Data Analysis 71*, 159–182.

Ingrassia, S., S. C. Minotti, A. Punzo, and G. Vittadini (2015). The generalized linear mixed cluster-weighted model. *Journal of Classification 32*(1), 85–113.

Ingrassia, S., S. C. Minotti, and G. Vittadini (2012). Local statistical modeling via the cluster-weighted approach with elliptical distributions. *Journal of Classification 29*(3), 363–401.

Jajuga, K. and D. Papla (2006). Copula functions in model based clustering. In M. Spiliopoulou, R. Kruse, C. Borgelt, A. Nürnberger, and W. Gaul (Eds.), *From Data and Information Analysis to Knowledge Engineering*, Studies in Classification, Data Analysis, and Knowledge Organization, pp. 603–613. Berlin: Springer.

Jasra, A., C. C. Holmes, and D. A. Stephens (2005). Markov chain Monte Carlo methods and the label switching problem in Bayesian mixture modeling. *Journal of the Royal Statistical Society: Series B 10*(1), 50–67.

Johnson, L. A. S. (1968a). Rainbow's end: The quest for an optimal taxonomy. *Proceedings of the Linnaean Society of New South Wales 93*, 8–45.

Johnson, L. A. S. (1970). Rainbow's end: The quest for an optimal taxonomy. *Systematic Biology 19*(3), 203–239.

Johnson, S. C. (1968b). Metric clustering. Unpublished manuscript. AT&T Bell Laboratories, Murray Hill, NJ.

Jordan, M. I., Z. Ghahramani, T. S. Jaakkola, and L. K. Saul (1999). An introduction to variational methods for graphical models. *Machine Learning 37*, 183–233.

Jöreskog, K. G. (1990). New developments in LISREL: Analysis of ordinal variables using polychoric correlations and weighted least squares. *Quality and Quantity 24*(4), 387–404.

Jørgensen, B. (1982). *Statistical Properties of the Generalized Inverse Gaussian Distribution*. New York: Springer-Verlag.

Karlis, D. and A. Santourian (2009). Model-based clustering with non-elliptically contoured distributions. *Statistics and Computing 19*(1), 73–83.

Kass, R. E. and A. E. Raftery (1995). Bayes factors. *Journal of the American Statistical Association 90*(430), 773–795.

Keribin, C. (2000). Consistent estimation of the order of mixture models. *Sankhyā. The Indian Journal of Statistics. Series A 62*(1), 49–66.

Kharin, Y. (1996). *Robustness in Statistical Pattern Recognition*. Dordrecht: Kluwer.

Kiers, H. A. L. (2002). Setting up alternating least squares and iterative majorization algorithms for solving various matrix optimization problems. *Computational Statistics and Data Analysis 41*(1), 157–170.

Kosmidis, I. and D. Karlis (2015). Model-based clustering using copulas with applications. *Statistics and Computing*. To appear.

Kotz, S., T. J. Kozubowski, and K. Podgorski (2001). *The Laplace Distribution and Generalizations: A Revisit with Applications to Communications, Economics, Engineering, and Finance* (1st ed.). Boston: Burkhäuser.

Kotz, S. and S. Nadarajah (2004). *Multivariate t Distributions and Their Applications*. Cambridge: Cambridge University Press.

Lagrange, J. L. (1788). *Méchanique Analitique*. Paris: Chez le Veuve Desaint.

Landsman, Z. M. and E. A. Valdez (2003). Tail conditional expectations for elliptical distributions. *North American Actuarial Journal 7*(4), 55–71.

Lange, K., D. R. Hunter, and I. Yang (2000). Optimization transfer using surrogate objective functions. *Journal of Computational and Graphical Statistics 9*, 1–20.

Lawley, D. N. and A. E. Maxwell (1962). Factor analysis as a statistical method. *Journal of the Royal Statistical Society: Series D 12*(3), 209–229.

Lee, S. and G. J. McLachlan (2011). On the fitting of mixtures of multivariate skew *t*-distributions via the EM algorithm. arXiv preprint arXiv:1109.4706.

Lee, S. and G. J. McLachlan (2014). Finite mixtures of multivariate skew t-distributions: some recent and new results. *Statistics and Computing 24*, 181–202.

Lee, S. X. and G. J. McLachlan (2013a). Model-based clustering and classification with non-normal mixture distributions. *Statistical Methods and Applications 22*(4), 427–454.

Lee, S. X. and G. J. McLachlan (2013b). On mixtures of skew normal and skew t-distributions. *Advances in Data Analysis and Classification 7*(3), 241–266.

Lefkomtch, L. P. (2004). Consensus principal components. *Biometrical Journal 35*, 567–580.

Leisch, F. (2004). Flexmix: A general framework for finite mixture models and latent class regression in R. *Journal of Statistical Software 11*(8), 1–18.

Leroux, B. G. (1992). Consistent estimation of a mixing distribution. *The Annals of Statistics 20*(3), 1350–1360.

Li, J. (2005). Clustering based on a multi-layer mixture model. *Journal of Computational and Graphical Statistics 14*(3), 547–568.

Li, K. C. (1991). Sliced inverse regression for dimension reduction (with discussion). *Journal of the American Statistical Association 86*, 316–342.

Li, K. C. (2000). High dimensional data analysis via the SIR/PHD approach. Unpublished manuscript.

Lichman, M. (2013). UCI machine learning repository. University of California, Irvine, School of Information and Computer Sciences.

Lin, T.-I. (2009). Maximum likelihood estimation for multivariate skew normal mixture models. *Journal of Multivariate Analysis 100*, 257–265.

Lin, T.-I. (2010). Robust mixture modeling using multivariate skew t distributions. *Statistics and Computing 20*(3), 343–356.

Lin, T.-I., G. J. McLachlan, and S. X. Lee (2016). Extending mixtures of factor models using the restricted multivariate skew-normal distribution. *Journal of Multivariate Analysis 143*, 398–413.

Lin, T.-I., P. D. McNicholas, and J. H. Hsiu (2014). Capturing patterns via parsimonious t mixture models. *Statistics and Probability Letters 88*, 80–87.

Lindsay, B. G. (1995). Mixture models: Theory, geometry and applications. In *NSF-CBMS Regional Conference Series in Probability and Statistics*, Volume 5. Hayward, CA: Institute of Mathematical Statistics.

Lindsey, J. K. (1999). Multivariate elliptically contoured distributions for repeated measurements. *Biometrics 55*(4), 1277–1280.

Lopes, H. F. and M. West (2004). Bayesian model assessment in factor analysis. *Statistica Sinica 14*, 41–67.

Lütkepohl, H. (1996). *Handbook of Matrices*. Chicester: John Wiley & Sons.

MacQueen, J. (1967). Some methods for classification and analysis of multivariate observations. In *Proceedings of the Fifth Berkeley Symposium on Mathematical Statistics and Probability, Volume 1: Statistics*, Berkeley, pp. 281–297. University of California Press.

Madigan, D. and A. E. Raftery (1994). Model selection and accounting for model uncertainty in graphical models using Occam's window. *Journal of the American Statistical Association 89*, 1535–1546.

Marbac, M., C. Biernacki, and V. Vandewalle (2014). Finite mixture model of conditional dependencies modes to cluster categorical data. arXiv preprint arXiv:1402.5103.

Marbac, M., C. Biernacki, and V. Vandewalle (2015). Model-based clustering of Gaussian copulas for mixed data. arXiv preprint arXiv:1405.1299v3.

Marbac, M. and M. Sedki (2015). Variable selection for model-based clustering using the integrated complete-data likelihood. arXiv preprint arXiv:1501.06314

Mardia, K. V., J. T. Kent, and J. M. Bibby (1980). *Multivariate Analysis*. Probability and Mathematical Statistics. London: Academic Press.

Markatou, M. (2000). Mixture models, robustness, and the weighted likelihood methodology. *Biometrics 56*(2), 483–486.

Marriott, F. H. C. (1974). *The Interpretation of Multiple Observations*. London: Academic Press.

Maugis, C. (2009). The selvarclust software. `www.math.univ-toulouse.fr/~maugis/SelvarClustHomepage.html`.

Maugis, C., G. Celeux, and M.–L. Martin–Magniette (2009a). Variable selection for clustering with Gaussian mixture models. *Biometrics 65*(3), 701–709.

Maugis, C., G. Celeux, and M.–L. Martin–Magniette (2009b). Variable selection in model-based clustering: A general variable role modeling. *Computational Statistics and Data Analysis 53*(11), 3872–3882.

McGrory, C. and D. Titterington (2007). Variational approximations in Bayesian model selection for finite mixture distributions. *Computational Statistics and Data Analysis 51*(11), 5352–5367.

McLachlan, G. J. (1992). *Discriminant Analysis and Statistical Pattern Recognition*. Hoboken, NJ: John Wiley & Sons.

McLachlan, G. J. and K. E. Basford (1988). *Mixture Models: Inference and Applications to Clustering*. New York: Marcel Dekker Inc.

McLachlan, G. J., R. W. Bean, and L. B.–T. Jones (2007). Extension of the mixture of factor analyzers model to incorporate the multivariate t-distribution. *Computational Statistics and Data Analysis 51*(11), 5327–5338.

McLachlan, G. J., R. W. Bean, and D. Peel (2002). A mixture model-based approach to the clustering of microarray expression data. *Bioinformatics 18*(3), 412–422.

McLachlan, G. J. and D. Peel (1998). Robust cluster analysis via mixtures of multivariate t-distributions. In *Lecture Notes in Computer Science*, Volume 1451, pp. 658–666. Berlin: Springer-Verlag.

McLachlan, G. J. and D. Peel (2000a). *Finite Mixture Models*. New York: John Wiley & Sons.

McLachlan, G. J. and D. Peel (2000b). Mixtures of factor analyzers. In *Proceedings of the Seventh International Conference on Machine Learning*, pp. 599–606. San Francisco: Morgan Kaufmann.

McNeil, A. J., R. Frey, and P. Embrechts (2005). *Quantitative Risk Management: Concepts, Techniques and Tools*. Princeton: Princeton University Press.

McNicholas, P. D. (2010). Model-based classification using latent Gaussian mixture models. *Journal of Statistical Planning and Inference 140*(5), 1175–1181.

McNicholas, P. D. and R. P. Browne (2013). Discussion of 'How to find an appropriate clustering for mixed-type variables with application to socio-economic stratification' by Hennig and Liao. *Journal of the Royal Statistical Society: Series C 62*(3), 352–353.

McNicholas, P. D., A. ElSherbiny, A. F. McDaid, and T. B. Murphy (2015). *pgmm: Parsimonious Gaussian Mixture Models*. R package version 1.2.

McNicholas, P. D., K. R. Jampani, and S. Subedi (2015). *longclust: Model-Based Clustering and Classification for Longitudinal Data*. R package version 1.2.

McNicholas, P. D. and T. B. Murphy (2005). Parsimonious Gaussian mixture models. Technical Report 05/11, Department of Statistics, Trinity College Dublin, Dublin, Ireland.

McNicholas, P. D. and T. B. Murphy (2008). Parsimonious Gaussian mixture models. *Statistics and Computing 18*(3), 285–296.

McNicholas, P. D. and T. B. Murphy (2010a). Model-based clustering of longitudinal data. *The Canadian Journal of Statistics 38*(1), 153–168.

McNicholas, P. D. and T. B. Murphy (2010b). Model-based clustering of microarray expression data via latent Gaussian mixture models. *Bioinformatics 26*(21), 2705–2712.

McNicholas, P. D., T. B. Murphy, A. F. McDaid, and D. Frost (2010). Serial and parallel implementations of model-based clustering via parsimonious Gaussian mixture models. *Computational Statistics and Data Analysis 54*(3), 711–723.

McNicholas, P. D. and S. Subedi (2012). Clustering gene expression time course data using mixtures of multivariate t-distributions. *Journal of Statistical Planning and Inference 142*(5), 1114–1127.

McNicholas, S. M., P. D. McNicholas, and R. P. Browne (2014). Mixtures of variance-gamma distributions. arXiv preprint arXiv:1309.2695v2.

Melnykov, V. (2016). Model-based biclustering of clickstream data. *Computational Statistics and Data Analysis 93*, 31–45.

Meng, X.-L. and D. B. Rubin (1993). Maximum likelihood estimation via the ECM algorithm: a general framework. *Biometrika 80*, 267–278.

Meng, X.-L. and D. van Dyk (1997). The EM algorithm — An old folk song sung to a fast new tune (with discussion). *Journal of the Royal Statistical Society: Series B 59*(3), 511–567.

Meyer, D., E. Dimitriadou, K. Hornik, A. Weingessel, and F. Leisch (2014). *e1071: Misc Functions of the Department of Statistics (e1071), TU Wien.* R package version 1.6-3.

Mirkin, B. G. and L. B. Chernyi (1970). Measurement of the distance between distinct partitions of a finite set of objects. *Automation and Remote Control 31*, 786–792.

Montanari, A. and C. Viroli (2010). Heteroscedastic factor mixture analysis. *Statistical Modelling 10*(4), 441–460.

Montanelli, Jr., R. G. and L. G. Humphreys (1976). Latent roots of random data correlation matrices with squared multiple correlations on the diagonal: A Monte Carlo study. *Psychometrika 41*, 341–348.

Morey, L. and A. Agresti (1984). The measurement of classification agreement: An adjustment to the Rand statistic for chance agreement. *Educational and Psychological Measurement 44*, 33–37.

Morris, K. and P. D. McNicholas (2013). Dimension reduction for model-based clustering via mixtures of shifted asymmetric Laplace distributions. *Statistics and Probability Letters 83*(9), 2088–2093. Erratum: vol. 85 (2014), 168.

Morris, K. and P. D. McNicholas (2016). Clustering, classification, discriminant analysis, and dimension reduction via generalized hyperbolic mixtures. *Computational Statistics and Data Analysis 97*, 133–150.

Morris, K., P. D. McNicholas, A. Punzo, and R. P. Browne (2014). Robust asymmetric clustering. arXiv preprint arXiv:1402.6744v1.

Morris, K., P. D. McNicholas, and L. Scrucca (2013). Dimension reduction for model-based clustering via mixtures of multivariate t-distributions. *Advances in Data Analysis and Classification* 7(3), 321–338.

Murray, P. M., R. B. Browne, and P. D. McNicholas (2013). Mixtures of 'unrestricted' skew-t factor analyzers. arXiv preprint arXiv:1310.6224.

Murray, P. M., R. B. Browne, and P. D. McNicholas (2014a). Mixtures of skew-t factor analyzers. *Computational Statistics and Data Analysis* 77, 326–335.

Murray, P. M., P. D. McNicholas, and R. B. Browne (2014b). A mixture of common skew-t factor analyzers. *Stat* 3(1), 68–82.

Muthen, B. and T. Asparouhov (2006). Item response mixture modeling: Application to tobacco dependence criteria. *Addictive Behaviors* 31, 1050–1066.

Nakai, K. and M. Kanehisa (1991). Expert system for predicting protein localization sites in gram-negative bacteria. *Proteins* 11(2), 95–110.

Nakai, K. and M. Kanehisa (1992). A knowledge base for predicting protein localization sites in eukaryotic cells. *Genomics* 14(4), 897–911.

O'Hagan, A., T. B. Murphy, I. C. Gormley, P. D. McNicholas, and D. Karlis (2016). Clustering with the multivariate normal inverse Gaussian distribution. *Computational Statistics and Data Analysis* 93, 18–30.

Orchard, T. and M. A. Woodbury (1972). A missing information principle: Theory and applications. In L. M. Le Cam, J. Neyman, and E. L. Scott (Eds.), *Proceedings of the Sixth Berkeley Symposium on Mathematical Statistics and Probability, Volume 1: Theory of Statistics*, pp. 697–715. Berkeley: University of California Press.

Pan, J. and G. MacKenzie (2003). On modelling mean-covariance structures in longitudinal studies. *Biometrika* 90(1), 239–244.

Pascal, F., L. Bombrun, J.–Y. Tourneret, and Y. Berthoumieu (2013). Parameter estimation for multivariate generalized Gaussian distributions. *IEEE Transactions on Signal Processing* 61(23), 5960–5971.

Pearson, K. (1894). Contributions to the mathematical theory of evolution. *Philosophical Transactions of the Royal Society, Part A 185*, 71–110.

Peel, D. and G. J. McLachlan (2000). Robust mixture modelling using the t distribution. *Statistics and Computing* 10(4), 339–348.

Pourahmadi, M. (1999). Joint mean-covariance models with applications to longitudinal data: Unconstrained parameterisation. *Biometrika 86*(3), 677–690.

Pourahmadi, M. (2000). Maximum likelihood estimation of generalised linear models for multivariate normal covariance matrix. *Biometrika 87*(2), 425–435.

Pourahmadi, M., M. Daniels, and T. Park (2007). Simultaneous modelling of the Cholesky decomposition of several covariance matrices. *Journal of Multivariate Analysis 98*, 568–587.

Punzo, A. (2014). Flexible mixture modeling with the polynomial Gaussian cluster-weighted model. *Statistical Modelling 14*(3), 257–291.

Punzo, A., R. P. Browne, and P. D. McNicholas (2016). Hypothesis testing for mixture model selection. *Journal of Statistical Computation and Simulation 86*(14), 2797–2818.

Punzo, A. and S. Ingrassia (2015). Clustering bivariate mixed-type data via the cluster-weighted model. *Computational Statistics*. To appear.

Punzo, A., A. Mazza, and P. D. McNicholas (2015). *ContaminatedMixt: Model-Based Clustering and Classification with the Multivariate Contaminated Normal Distribution*. R package version 1.0.

Punzo, A. and P. D. McNicholas (2014a). Robust clustering in regression analysis via the contaminated Gaussian cluster-weighted model. arXiv preprint arXiv:1409.6019v1.

Punzo, A. and P. D. McNicholas (2014b). Robust high-dimensional modeling with the contaminated Gaussian distribution. arXiv preprint arXiv:1408.2128v1.

Punzo, A. and P. D. McNicholas (2016). Parsimonious mixtures of multivariate contaminated normal distributions. arXiv preprint arXiv:1305.4669v5.

R Core Team (2015). *R: A Language and Environment for Statistical Computing*. Vienna, Austria: R Foundation for Statistical Computing.

Raftery, A. E. and N. Dean (2006). Variable selection for model-based clustering. *Journal of the American Statistical Association 101*(473), 168–178.

Ranalli, M. and R. Rocci (1971). Mixture methods for ordinal data: A pairwise likelihood approach *Statistics and Computing 26*(1), 529–547.

Rand, W. M. (1971). Objective criteria for the evaluation of clustering methods. *Journal of the American Statistical Association 66*(336), 846–850.

Rao, C. R. (1952). *Advanced Statistical Methods in Biometric Research*. New York: John Wiley and Sons, Inc.

Rau, A., C. Maugis-Rabusseau, M.–L. Martin–Magniette, and G. Celeux (2015). Co-expression analysis of high-throughput transcriptome sequencing data with Poisson mixture models. *Bioinformatics 31*(9), 1420–1427.

Reaven, G. M. and R. G. Miller (1979). An attempt to define the nature of chemical diabetes using a multidimensional analysis. *Diabetologica 16*(1), 17–24.

Richardson, S. and P. J. Green (1997). On Bayesian analysis of mixtures with an unknown number of components (with discussion). *Journal of the Royal Statistical Society: Series B 59*(4), 731–792.

Ritter, G. (2015). *Robust Cluster Analysis and Variable Selection*, Volume 137 of *Chapman & Hall/CRC Monographs on Statistics & Applied Probability*. Boca Raton: Chapman & Hall/CRC Press.

Sahu, K., D. K. Dey, and M. D. Branco (2003). A new class of multivariate skew distributions with applications to Bayesian regression models. *Canadian Journal of Statistics 31*(2), 129–150. Corrigendum: vol. 37 (2009), 301–302.

Schöner, B. (2000). *Probabilistic Characterization and Synthesis of Complex Data Driven Systems*. Ph.D. thesis, MIT.

Schroeter, P., J. Vesin, T. Langenberger, and R. Meuli (1998). Robust parameter estimation of intensity distributions for brain magnetic resonance images. *IEEE Transactions on Medical Imaging 17*(2), 172–186.

Schwarz, G. (1978). Estimating the dimension of a model. *The Annals of Statistics 6*(2), 461–464.

Scott, A. J. and M. J. Symons (1971). Clustering methods based on likelihood ratio criteria. *Biometrics 27*, 387–397.

Scrucca, L. (2010). Dimension reduction for model-based clustering. *Statistics and Computing 20*(4), 471–484.

Scrucca, L. (2014). Graphical tools for model-based mixture discriminant analysis. *Advances in Data Analysis and Classification 8*(2), 147–165.

Spearman, C. (1904). The proof and measurement of association between two things. *American Journal of Psychology 15*, 72–101.

Spearman, C. (1927). *The Abilities of Man: Their Nature and Measurement*. London: MacMillan and Co., Limited.

Spiegelhalter, D. J., N. G. Best, B. P. Carlin, and A. Van der Linde (2002). Bayesian measures of model complexity and fit (with discussion). *Journal of the Royal Statistical Society: Series B 64*, 583–639.

Steane, M. A., P. D. McNicholas, and R. Yada (2012). Model-based classification via mixtures of multivariate t-factor analyzers. *Communications in Statistics – Simulation and Computation 41*(4), 510–523.

Steele, R. J. and A. E. Raftery (2010). Performance of Bayesian model selection criteria for Gaussian mixture models. In *Frontiers of Statistical Decision Making and Bayesian Analysis*, Volume 2, pp. 113–130. New York: Springer.

Steinley, D. (2004). Properties of the Hubert–Arabie adjusted Rand index. *Psychological Methods 9*, 386–396.

Steinley, D. (2006). K-means clustering: A half-century synthesis. *British Journal of Mathematical and Statistical Psychology 59*, 1–34.

Steinley, D. and M. J. Brusco (2008). Selection of variables in cluster analysis: An empirical comparison of eight procedures. *Psychometrika 73*, 125–144.

Steinley, D. and M. J. Brusco (2011a). Evaluating mixture modeling for clustering: Recommendations and cautions. *Psychological Methods 16*(1), 63–79.

Steinley, D. and M. J. Brusco (2011b). *K*-means clustering and mixture model clustering: Reply to McLachlan (2011) and Vermunt (2011) *Psychological Methods 16*(1), 89–92.

Stephens, M. (1997). *Bayesian Methods for Mixtures of Normal Distributions*. Ph.D. thesis, University of Oxford, Oxford, England.

Stephens, M. (2000). Bayesian analysis of mixture models with an unknown number of components — An alternative to reversible jump methods. *The Annals of Statistics 28*(1), 40–74.

Streuli, H. (1973). Der heutige stand der kaffeechemie. In *Association Scientifique International du Cafe, 6th International Colloquium on Coffee Chemisrty*, Bogatá, Columbia, pp. 61–72.

Subedi, S. and P. D. McNicholas (2014). Variational Bayes approximations for clustering via mixtures of normal inverse Gaussian distributions. *Advances in Data Analysis and Classification 8*(2), 167–193.

Subedi, S. and P. D. McNicholas (2016). A variational approximations-DIC rubric for parameter estimation and mixture model selection within a family setting. arXiv preprint arXiv:1306.5368v2.

Subedi, S., A. Punzo, S. Ingrassia, and P. D. McNicholas (2013). Clustering and classification via cluster-weighted factor analyzers. *Advances in Data Analysis and Classification 7*(1), 5–40.

Subedi, S., A. Punzo, S. Ingrassia, and P. D. McNicholas (2015). Cluster-weighted t-factor analyzers for robust model-based clustering and dimension reduction. *Statistical Methods and Applications 24*(4), 623–649.

Sundberg, R. (1974). Maximum likelihood theory for incomplete data from an exponential family. *Scandinavian Journal of Statistics 1*(2), 49–58.

Synge, J. L. (1951). *Science: Sense and Nonsense.* New York: W. W. Norton & Company, Inc.

Tang, Y., R. P. Browne, and P. D. McNicholas (2015). Model-based clustering of high-dimensional binary data. *Computational Statistics and Data Analysis 87*, 84–101.

Teschendorff, A., Y. Wang, J. Barbosa-Morais, N. Brenton, and C. Caldas (2005). A variational Bayesian mixture modelling framework for cluster analysis of gene-expression data. *Bioinformatics 21*(13), 3025–3033.

Thorndike, R. L. (1953). Who belongs in the family? *Psychometrika 18*, 267–276.

Tiedeman, D. V. (1955). On the study of types. In S. B. Sells (Ed.), *Symposium on Pattern Analysis*, pp. 1–14. Randolph Field, TX: Air University, U.S.A.F. School of Aviation Medicine.

Tipping, M. E. and C. M. Bishop (1997). Mixtures of probabilistic principal component analysers. Technical Report NCRG/97/003, Aston University (Neural Computing Research Group), Birmingham, UK.

Tipping, M. E. and C. M. Bishop (1999). Mixtures of probabilistic principal component analysers. *Neural Computation 11*(2), 443–482.

Titterington, D. M., A. F. M. Smith, and U. E. Makov (1985). *Statistical Analysis of Finite Mixture Distributions.* Chichester: John Wiley & Sons.

Tortora, C., R. P. Browne, B. C. Franczak, and P. D. McNicholas (2015a). *MixGHD: Model Based Clustering, Classification and Discriminant Analysis Using the Mixture of Generalized Hyperbolic Distributions.* R package version 1.8.

Tortora, C., B. C. Franczak, R. P. Browne, and P. D. McNicholas (2014). Mixtures of multiple scaled generalized hyperbolic distributions. arXiv preprint arXiv:1403.2332v2.

Tortora, C., B. C. Franczak, R. P. Browne, and P. D. McNicholas (2016). A mixture of coalesced generalized hyperbolic distributions. arXiv preprint arXiv:1403.2332v7.

Tortora, C., P. D. McNicholas, and R. P. Browne (2015b). A mixture of generalized hyperbolic factor analyzers. *Advances in Data Analysis and Classification.* To appear.

Van Deun, K., T. F. Wilderjans, R. A. Van Den Berg, A. Antoniadis, and I. Van Mechelen (2011). A flexible framework for sparse simultaneous component based data integration. *BMC Bioinformatics 12*, 1971–1987.

Verdoolaege, G., S. De Backer, and P. Scheunders (2008). Multiscale colour texture retrieval using the geodesic distance between multivariate generalized Gaussian models. In *15th IEEE International Conference on Image Processing, 2008. ICIP 2008*, pp. 169–172.

Vermunt, J. K. (2003). Multilevel latent class models. *Sociological Methodology 33*(1), 213–239.

Vermunt, J. K. (2007). Multilevel mixture item response theory models: An application in education testing. In *Proceedings of the 56th Session of the International Statistical Institute*, Lisbon, Portugal, pp. 22–28.

Vermunt, J. K. (2011). *K*-means may perform as well as mixture model clustering but may also be much worse: Comment on Steinley and Brusco (2011). *Psychological Methods 16*(1), 82–88.

Vrac, M., L. Billard, E. Diday, and A. Chedin (2012). Copula analysis of mixture models. *Computational Statistics 27*(3), 427–457.

Vrbik, I. and P. D. McNicholas (2012). Analytic calculations for the EM algorithm for multivariate skew-t mixture models. *Statistics and Probability Letters 82*(6), 1169–1174.

Vrbik, I. and P. D. McNicholas (2014). Parsimonious skew mixture models for model-based clustering and classification. *Computational Statistics and Data Analysis 71*, 196–210.

Vrbik, I. and P. D. McNicholas (2015). Fractionally-supervised classification. *Journal of Classification 32*(3), 359–381.

Wang, Q., C. Carvalho, J. Lucas, and M. West (2007). BFRM: Bayesian factor regression modelling. *Bulletin of the International Society for Bayesian Analysis 14*(2), 4–5.

Waterhouse, S., D. MacKay, and T. Robinson (1996). Bayesian methods for mixture of experts. In *Advances in Neural Information Processing Systems*, Volume 8. Cambridge, MA: MIT Press.

Wei, Y. and P. D. McNicholas (2015). Mixture model averaging for clustering. *Advances in Data Analysis and Classification 9*(2), 197–217.

Weisberg, S. (2005). *Applied Linear Regression* (Third ed.). Hoboken, NJ: Wiley.

Weisberg, S. (2010). *alr3: Companion to Applied Linear Regression*. R package version 2.0.

West, M. (2003). Bayesian factor regression models in the "large p, small n" paradigm. In J. M. Bernardo, M. Bayarri, J. Berger, A. Dawid, D. Heckerman, A. Smith, and M. West (Eds.), *Bayesian Statistics*, Volume 7, Oxford, pp. 723–732. Oxford University Press.

Wickham, H. (2009). *ggplot2: Elegant Graphics for Data Analysis*. New York: Springer.

Wolfe, J. H. (1963). *Object Cluster Analysis of Social Areas*. Master's thesis, University of California, Berkeley.

Wolfe, J. H. (1965). A computer program for the maximum likelihood analysis of types. Technical Bulletin 65-15, U.S. Naval Personnel Research Activity.

Wolfe, J. H. (1970). Pattern clustering by multivariate mixture analysis. *Multivariate Behavioral Research 5*, 329–350.

Woodbury, M. A. (1950). *Inverting modified matrices*. Statistical Research Group, Memorandum Report 42. Princeton, NJ: Princeton University.

Wraith, D. and F. Forbes (2014). Clustering using skewed multivariate heavy tailed distributions with flexible tail behaviour. arXiv preprint arXiv:1408.0711.

Yoshida, R., T. Higuchi, and S. Imoto (2004). A mixed factors model for dimension reduction and extraction of a group structure in gene expression data. In *Proceedings of the 2004 IEEE Computational Systems Bioinformatics Conference*, pp. 161–172.

Yoshida, R., T. Higuchi, S. Imoto, and S. Miyano (2006). ArrayCluster: An analytic tool for clustering, data visualization and module finder on gene expression profiles. *Bioinformatics 22*, 1538–1539.

Zhang, J. and F. Liang (2010). Robust clustering using exponential power mixtures. *Biometrics 66*(4), 1078–1086.

Zhou, H. and K. L. Lange (2010). On the bumpy road to the dominant mode. *Scandinavian Journal of Statistics 37*(4), 612–631.

Index